The Weighted Airman Promotion System

Standardizing Test Scores

Michael Schiefer, Albert A. Robbert, John S. Crown,
Thomas Manacapilli, Carolyn Wong

Prepared for the United States Air Force
Approved for public release; distribution unlimited

RAND PROJECT AIR FORCE

The research described in this report was sponsored by the United States Air Force under Contract FA7014-06-C-0001. Further information may be obtained from the Strategic Planning Division, Directorate of Plans, Hq USAF.

Library of Congress Cataloging-in-Publication Data

The weighted airman promotion system : standardizing test scores / Michael Schiefer ... [et al.].
p. cm.
Includes bibliographical references.
ISBN 978-0-8330-4231-6 (pbk. : alk. paper)
1. United States. Air Force—Personnel management. 2. United States. Air Force—Airmen—Promotions. 3. United States. Air Force—Airmen—Ability testing. I. Schiefer, Michael.

UG793.W45 2008
358.4'11120973—dc22

2008018789

Published 2008 by the RAND Corporation
1776 Main Street, P.O. Box 2138, Santa Monica, CA 90407-2138
1200 South Hayes Street, Arlington, VA 22202-5050
4570 Fifth Avenue, Suite 600, Pittsburgh, PA 15213-2665
RAND URL: http://www.rand.org/
To order RAND documents or to obtain additional information, contact
Distribution Services: Telephone: (310) 451-7002;
Fax: (310) 451-6915; Email: order@rand.org

Preface

This study, conducted in the Manpower, Personnel, and Training Program of RAND Project AIR FORCE (PAF), is a follow-on to *Air Force Enlisted Force Management: System Interactions and Synchronization Strategies* (Schiefer et al., 2007). This monograph explores a potential modification to the enlisted promotion system, one of the primary systems that affect the enlisted force.

Brig Gen Glenn Spears sponsored this work in fiscal year 2006 as the Director of Force Management Policy, Deputy Chief of Staff for Personnel, Headquarters U.S. Air Force. The research was accomplished as part of a project entitled Enlisted Force Management. This monograph should be of interest to those responsible for Air Force enlisted testing and promotion policies, to those who develop enlisted promotion tests, to strength managers, to the Chief's Group at the Air Force Personnel Center, and to enlisted career field managers.

We appreciate that the findings in this monograph will not generate change without the support of key leaders within the Air Force. The document was prepared with that audience in mind.

RAND Project Air Force

RAND Project AIR FORCE, a division of the RAND Corporation, is the U.S. Air Force's federally funded research and development center for studies and analyses. PAF provides the Air Force with independent analyses of policy alternatives affecting the development, employment, combat readiness, and support of current and future aerospace forces.

Research is conducted in four programs: Aerospace Force Development; Manpower, Personnel, and Training; Resource Management; and Strategy and Doctrine.

Additional information about PAF is available on our Web site: http://www.rand.org/paf/

Contents

Figures

Tables

Summary

The U.S. Air Force has three major independent systems that affect the health of its enlisted force: the manpower system, the strength management system, and the enlisted promotion system. Because the current organizational structure lacks broad coordinating and control mechanisms, this independence spawns policies and procedures that occasionally work at cross-purposes. We discuss these systems at length in *Air Force Enlisted Force Management: System Interactions and Synchronization Strategies* (Schiefer et al., 2007). That monograph proposes multiple follow-on efforts, and this study fulfills one of those recommendations.

Specifically, we examine the practice of not standardizing the test scores that are part of the enlisted promotion system.[1] This practice produces results that are inconsistent with two overarching policies. First, Air Force Policy Directive 36-25 requires that the enlisted promotion system "identify those people with the highest potential to fill positions of increased grade and responsibility."[2] We show that not standardizing test scores means that the Air Force emphasizes longevity and testing ability differently across and within specialties to identify individuals

---

1 Many, if not most, tests that are administered to different groups at different times are standardized. Standardization involves mathematically transforming raw test scores into new scores with desirable properties. For example, the Armed Forces Qualification Test (AFQT) reports standardized scores, so that an AFQT score of 72 represents the same level of ability today as it did four years ago. Were it not for standardized scores, the military services could not track the quality of new recruits over time.

2 U.S. Air Force, 1993, p. 1.

with the highest potential. Further, we demonstrate that these standards vary randomly over time. Random variations in the impacts of selection criteria make it difficult to understand how the Air Force can be achieving its primary promotion policy objective.

Our second concern deals with differences in promotion opportunity. While the testing dimension of the enlisted promotion system allows members to influence their own destinies, not standardizing scores means that members of specialties in which testing carries more weight have more control than members of other specialties do. This produces random promotion opportunity differences across Air Force specialty codes (AFSCs), thus violating an equity principle that can be traced to a 1970s-era strategic plan for enlisted force management known as the Total Objective Plan for Career Airman Personnel (TOPCAP).[3] Because the Air Force does not standardize test scores, the current policy of equal *selection* opportunity does not imply equal *promotion* opportunity over a career. Consequently, there is a greater opportunity to achieve senior enlisted grades in some AFSCs than in others.

The random aspects of the enlisted promotion system also produce other potentially undesirable consequences. For example, not standardizing scores yields unpredictable manning percentages by specialty. This has negative force management implications. Uncertainty also means that the Air Force, when it fills future strategic chief master sergeant positions, will disproportionately draw from specialties in which testing carries more weight.[4]

The modification we propose would not change equal selection opportunity. However, it would affect selection decisions within AFSCs. Test score standardization would primarily affect those com-

---

[3] The Air Force Personnel Plan (U.S. Air Force, 1975) provides TOPCAP details. A primary objective of TOPCAP was to maintain a career force, and it established a promotion system founded on equity across specialties. That culture of equity persists throughout the enlisted force today, and subsequent personnel plans have consistently stressed the importance of equity. One premise of TOPCAP was that promotion equity and predictability were keys to realizing retention rates that would sustain the career enlisted force.

[4] The Air Force fills strategic chief positions through commander involvement or nomination processes.

peting for selection to E5–E7. It would have extremely limited impacts on E8 and E9 selections, which are determined primarily by selection board scores.

After presenting supporting data, we discuss a range of outcomes that the Air Force could achieve by adopting various standardization strategies. We recommend that the Air Force leadership implement a standardization strategy that will produce predictable outcomes that are consistent with its personnel priorities and policies.

Acknowledgments

We could not have initiated this work without the sponsorship of Brig Gen Glenn Spears, who provided Project AIR FORCE with the opportunity to reengage in enlisted management issues. We thank John Park, Tina Strickland, and Lisa Mills from the Deputy Chief of Staff for Personnel's (AF/A1) Force Policy Management Division, Gwen Rutherford from the Leadership Transformation and Integration Division, and CMSgt Trenda Voegtle from the Promotion Policy Division for their insights. CMSgt Rusty Nicholson, Ken Schwartz, and Johnny Weissmuller at the Air Force Personnel Center generously shared their knowledge of the enlisted promotion system.

We are also grateful to Lt Col Jim Wisnowski, Commander of the Air Force Occupational Measurement Squadron. In addition, we thank Julie Duminiak and Neil Dorans from the Educational Testing Service (ETS) who directed us to ETS studies and responded to our inquiries regarding ETS's approach to standardization. We thank Jim Sconing for his detailed explanations of standardization for the American College Test (ACT).

We thank our colleagues Harry Thie and John Drew and Jay Jacobson (Air Force retired) for their extremely constructive formal reviews. Finally, we wish to acknowledge our editor, Miriam Polon.

Abbreviations

ACT	American College Test
AF/A1	Air Force Deputy Chief of Staff for Manpower and Personnel
AFI	Air Force Instruction
AFQT	Armed Forces Qualification Test
AFPC	Air Force Personnel Center
AFPC/DPP	AFPC Directorate of Personnel Programs
AFSC	Air Force specialty code
ASVAB	Armed Services Vocational Aptitude Battery
CAREERS	Career Airman Reenlistment Reservation System
CCS	chronic critical shortage
CJR	career job reservation
DoD	Department of Defense
DoDD	Department of Defense Directive
E1	airman basic
E2	airman
E3	airman first class
E4	senior airman
E5	staff sergeant
E6	technical sergeant
E7	master sergeant
E8	senior master sergeant
E9	chief master sergeant
EPR	enlisted performance report

ESO	equal selection opportunity
ETS	Educational Testing Service
FB	fast burner
FY	fiscal year
HYT	high year of tenure
IDEAS	AFPC's Interactive Demographic Analysis System
IEB	initial enlistment bonus
MAGE	mechanical, administrative, general, electronic
NCO	noncommissioned officer
NPS	non–prior service
OSD	Office of the Secretary of Defense
OSI	Office of Special Investigations
PFE	Promotion Fitness Exam
RAW	Retrieval Application Website (AFPC)
ROTC	Reserve Officer Training Corps
SAT	Scholastic Aptitude Test
SECAF	Secretary of the Air Force
SKT	Specialty Knowledge Test
SRB	selective reenlistment bonus
TIG	time in grade
TIS	time in service
TOPCAP	Total Objective Plan for Career Airman Personnel
TPR	trained personnel requirement
UIF	unfavorable information file
WAPS	Weighted Airman Promotion System
YOS	years of service

CHAPTER ONE

Introduction

This monograph is an extension of *Air Force Enlisted Force Management: System Interactions and Synchronization Strategies* (Schiefer et al., 2007), which discusses policy options that the Air Force employs in its efforts to manage the active-duty enlisted force. One of the main messages of the earlier study is that enlisted strength managers need to better synchronize the three primary control systems that affect the health of the enlisted force:

- the strength management system, which establishes targets for total strength, recruiting, retraining, and bonuses
- the manpower system, which sets requirements for each grade and specialty combination in the form of authorizations
- the enlisted promotion system, which determines the annual number of selections by grade in the aggregate and in each specialty.[1]

The Air Force currently tends to manage these systems in isolation. However, actions taken to control one system often affect another. For example, the earlier study postulated that the Air Force's policy of not standardizing the test scores that are part of the enlisted promotion system might be having adverse impacts on the strength manage-

[1] In this monograph, *aggregate* means all specialties considered as a group. *Disaggregate* means at the Air Force specialty code (AFSC) level of detail.

ment system.[2] In this monograph, we demonstrate that these theoretical effects are real. We also surface a more pressing issue. We believe that not standardizing test scores means that the enlisted promotion system cannot be achieving its primary objective.

The Air Force Cannot Be Achieving Its Primary Enlisted Promotion Objective

The fundamental principle that governs enlisted promotions today is stated in Air Force Policy Directive 36-25:

> 1. The Air Force must be able to identify those people with the highest potential to fill positions of increased grade and responsibility.[3]

The Air Force does not explicitly define what it means by "highest potential." Air Force Instruction 36-2502 does specify that

---

2 Standardizing scores would involve mathematically converting raw scores into adjusted scores that have desirable properties (see Chapter Three). For example, testing experts might wish to standardize test scores so that the score distribution across all test-takers in every AFSC had the same bell-shaped curve from one year to the next. Standardizing test scores would not change each individual's testing rank order within an AFSC relative to other testers. However, it would usually change the differences between each individual's test scores and the scores of other testers. Hence, standardized test scores, when combined with the points from other factors of the Weighted Airman Promotion System (WAPS, see Chapter Two), would partially modify the selection list for an AFSC. However, the number of promotion selections within each AFSC would remain unchanged under the policy of equal selection opportunity (ESO).

As originally implemented, WAPS test scores were based on a percentile ranking. Hence, scores ranged from 0 to 100 on both the Promotion Fitness Exam (PFE) and Specialty Knowledge Test (SKT). Using a percentile ranking was one way to standardize because every AFSC had approximately the same distribution of test scores. However, because missing one additional question could substantially change one's percentile ranking in a large AFSC with many tie scores, the Air Force started basing test scores on the percentage of correct answers in 1972. With this change, scores were no longer standardized, and some of the Air Force's original promotion equity objectives slipped from reach—although there is no evidence to suggest that anyone realized at the time that there was a connection between achieving promotion equity and standardizing test scores.

3 U.S. Air Force, 1993, p. 1.

the Weighted Airman Promotion System (WAPS) is the mechanism through which the Air Force promotes individuals and that each AFSC[4] will have an equal selection opportunity (ESO):[5]

> 2.3. SSgt, TSgt, or MSgt Promotions . . . Airmen compete and test under the Weighted Airman Promotion System (WAPS) in the Control Air Force Specialty Code (CAFSC) held on the PECD [promotion eligibility cutoff date] . . .
>
> 2.3.1. HQ AFPC/DPPPWM:
>
> 2.3.1.1. Makes promotion selections' using the WAPS and data in PDS [the Personnel Data System].
>
> 2.3.1.3. Makes promotion selections by computer, applies the quota equally to each promotion Air Force Specialty Code (AFSC), and ensures equal selection opportunity (ESO) for all AFSCs.[6]

Presumably, WAPS should be designed to promote those with the highest potential. However, we show that not standardizing the test scores that are part of WAPS means that the Air Force emphasizes longevity and specialty knowledge (as measured by testing) differently across and within specialties when it identifies the highest potential individuals. Further, we demonstrate that these standards vary randomly over time. Therefore, WAPS is not producing deliberate and consistent results. Hence, we are not persuaded that the enlisted promotion system is achieving its primary policy objective.

While the impacts of not standardizing test scores are not common knowledge, there is almost universal agreement that enlisted

---

4 In this monograph, *AFSC* means *promote-to AFSC*. Members compete for selection in promote-to AFSCs, which normally correspond to *control AFSCs* (CAFSCs).

5 The ESO policy allows a measured departure from this overall scheme for mission-critical AFSCs with *chronic critical shortages* (CCS). These AFSCs realize selection rates that are 1.2 times the rates in other AFSCs. In March 2003, the Air Force enhanced the CCS program for E8 and E9 cycles by promoting some AFSCs below the AF average to free up selections for the CCS skills. The Air Force also promotes some to E8 and E9 in CCSs at greater than 1.2 times the Air Force rate.

6 U.S. Air Force, 2002, p. 15.

promotion-system policies that strive for equity disrupt the disaggregate strength management system. As background material, we discuss the evolution of the Air Force's enlisted culture of promotion equity.

The Motivation for Enlisted Promotion Equity

Air Force enlisted promotion policies emphasize equity across AFSCs. Striving for promotion equity directly magnifies, but may indirectly mitigate, manning deviations. The Air Force once believed that promotion equity and predictability were the keys to high retention rates and that good retention was the long-term key to sustaining the enlisted career force in the aggregate. Hence, promotion equity was intended to ensure that there was at least an aggregate base of experienced enlisted members from which to operate. The emphasis on the equity-predictability-retention relationship can be directly traced to the Total Objective Plan for Career Airman Personnel (TOPCAP), which the Air Force implemented in the early 1970s:

> 1-3 b. A fundamental aim of TOPCAP is to build a career plan that will influence adequate numbers of airmen to elect career status.
>
> 1-3 d. Airmen desire a high order of stability and consistency in personnel policies that affect them. Their entitlement to career visibility and equitable consideration in programs, policies, and objectives must be recognized and emphasized.
>
> 3-6 e. TOPCAP guarantees promotion opportunity to each grade as follows:
>
> 90 percent to staff sergeant
> 90 percent to technical sergeant
> 84 percent to master sergeant
> 75 percent to senior master sergeant
> 60 percent to chief master sergeant
>
> TOPCAP thus tells the individual airman when he can expect to be promoted, what his chances of being promoted are, how

he will be selected, and what the consequences are of not being promoted.[7]

TOPCAP's guarantee of equal promotion opportunity, in the presence of high year of tenure constraints, meant that promotion timing had to be the same, or nearly so, for every AFSC.[8] It also meant that the Air Force had to adopt a policy of equal selection opportunity:[9]

> 3-7 e. Promotion Management. Current promotion eligibility criteria, except for the grade/skill relationship, will be used in conjunction with the Weighted Airman Promotion System to select airmen for promotion. Equal selection opportunity will be provided each airman through the promotion zones without regard to AFSC.[10]

Hall and Nelsen note that, from the outset, the Air Force realized that ESO would disrupt disaggregate manning:

> The basic principle of the TOPCAP promotion plan was to provide equal selection opportunity for all airmen regardless of AFSC. This principle also states that the average time for promotion to each grade should be the same for each AFSC. Under the equal selection concept, which was adopted in July, 1972, each competing AFSC received an equal percentage of the overall promotion quota without regard to manning. This represented a

---

[7] U.S. Air Force, 1975, pp. 1-1 to 3-2.

[8] *High year of tenure* (HYT) is a policy that limits the number of years that enlisted members can remain in the Air Force as a function of grade. The principal motivations for HYT are to keep selection rates higher and phase points lower by separating those who have fallen behind their peers (*phase points* are the years of service that members have when they are promoted).

[9] While *selection* and *promotion* are often used interchangeably, in this document we use *selection* in association with single annual cycles in which individuals compete for advancement in grade. We use *promotion* in the context of multiple selection cycles. Hence, *selection opportunity* means the percent of eligibles who were selected to be advanced in grade during a single cycle. *Promotion opportunity* is the probability of being identified for advancement over a career.

[10] U.S. Air Force, 1975. p. 3-7.

> major change in promotion philosophy from the previous system which allocated promotion quotas on the basis of AFSC vacancies via the promotion management list (PML). As was mentioned in Chapter III, use of the PML in the airman promotion system remained an area of criticism after the WAPS implementation. However, with the adoption of the equal selection concept, this area of criticism was removed. TOPCAP's equal selection concept became popular with the enlisted force because no AFSC was closed for promotion since each AFSC received the same promotion opportunity.
>
> However, since promotions were no longer being made to fill AFSC grade vacancies, the problem of grade imbalances continued. This was because equal selection opportunity aggravated surplus conditions and often did not supply enough promotions to fill a shortage condition. However, the value of equal selection opportunity was considered sufficiently great to offset its adverse effects on grade imbalances. The Air Force's position was that the promotion programs should not be used to solve manning imbalances but should be used to advance airmen who demonstrated potential for increased responsibility by means of objective and visible systems.[11]

To better appreciate the Air Force's preoccupation with sustaining the aggregate career force in the mid-1970s, it is helpful to recall that these were the early years of the all-volunteer force—an untested concept—and that the post-Vietnam drawdown was spawning what would ultimately become known as the "hollow force." However, because the Air Force has experienced a steady decline in its enlisted strength (a 43 percent reduction from 479,585 in 1976 to 273,990 in 2006) and because the Air Force achieved this reduction primarily by reducing accessions, experience levels have remained high and aggregate enlisted retention has not been a chronic issue.

---

[11] Hall and Nelsen, 1980, pp. 70–71.

The Air Force Has Not Achieved All TOPCAP and Secondary Promotion Objectives

The Air Force also prescribes secondary objectives for its enlisted promotion system that reflect three of the original TOPCAP promotion objectives:

> The enlisted promotion system supports DoDD 1304.20, Enlisted Personnel Management System, by helping to provide a visible, relatively stable career progression opportunity over the long term; attracting, retaining, and motivating to career service the kinds and numbers of people the military services need; and ensuring a reasonably uniform application of the principle of equal pay for equal work among the military services.[12]

While ESO has had the predicted adverse impacts on disaggregate manning, as implemented, it has not achieved all its secondary and TOPCAP equity objectives. To illustrate one shortfall, the horizontal axis in Figure 1.1 shows the ratio of the sum of the E7, E8, and E9 inventories to E6s[13] by 2-digit, non-tax[14] AFSC (see Appendix B for a list of AFSCs that make up each 2-digit grouping). In some AFSCs, the inventory in the top three grades is only about 50 percent

---

[12] U.S. Air Force, 2005a, p. 268.

[13] In the Air Force, E1 = airman basic, E2 = airman, E3 = airman first class, E4 = senior airman, E5 = staff sergeant, E6 = technical sergeant, E7 = master sergeant, E8 = senior master sergeant, E9 = chief master sergeant. E7s–E9s (top three) are *senior noncommissioned officers* (SNCOs).

[14] The Air Force uses five characters codes to identify AFSCs. These characters were digits in the predecessor to the current system, hence the moniker "2-digit." Today, the first character may have a value of 1–9 to designate a broad functional category. AFSCs that begin with 8 or 9 designate special-duty and reporting identifiers and are commonly referred to as *tax AFSCs* because they draw their inventories from AFSCs that begin with 1 through 7. The second position is a letter that designates a subgroup of specialties within the broad category. The third and fifth positions are numbers that identify specific specialties within the second position subgroup. The fourth position of the AFSC is the *skill level* (1-level= input, 3-level = apprentice, 5-level = journeyman, 7-level = craftsman, 9-level = superintendent, and 0-level = chief enlisted manager.) When an "X" is used as a placeholder, we mean all valid values for that position in the AFSC designation.

of the E6 population. In other AFSCs, the ratio is over 100 percent—reflecting a potential career promotion opportunity advantage. One would hope that the differences in inventory ratios would be driven by other programs that compensated for ESO in order to satisfy manning requirements. However, that is not the case. The vertical axis in Figure 1.1 shows that the top three/E6 requirements ratios range from about 50 percent to 125 percent. However, the shotgun-blast pattern in the figure indicates that there is little relationship between requirements ratios and inventory ratios. For example, the AFSCs captured by the oval all have top three/E6 requirements ratios of about 80 percent. However, the inventory ratios in these AFSCs range from 50 percent to 90 percent. The largest AFSC, Security Forces (3P), had a top three/E6

Figure 1.1
Disparate Seniority Ratios, September 2006

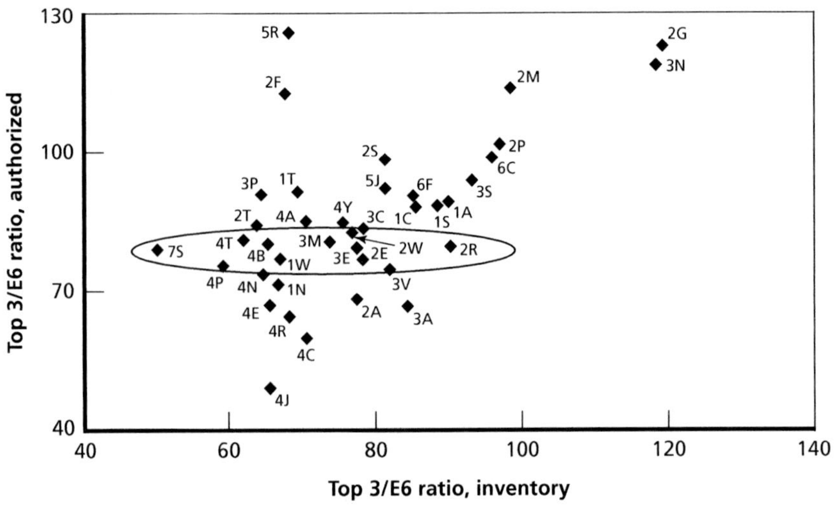

SOURCE: Derived from the Air Force Personnel Center's (AFPC's) Retrieval Applications Website (RAW).
NOTE: Represents non-tax, 2-digit, duty AFSCs with 50 or more top 3 assigned.
RAND *MG678-1.1*

requirement of 91 percent, but its top three/E6 inventory was only 64 percent.[15]

While retention differences beyond 20 years of service (YOS) might account for some of the inventory-to-requirements mismatch, in this monograph we make the case that not standardizing test scores is a major contributing factor. Thus, we believe that individuals are much more likely to achieve top-three status in some AFSCs than in others.

Because the Air Force spends hundreds of millions of dollars per year for retraining and reenlistment bonuses to counter the negative disaggregate aspects of ESO, we conclude that adhering to TOPCAP's equity philosophy is still very important, if for no other reason than equity has become a pillar of the enlisted culture. However, the Air Force does not actually achieve equal promotion opportunity across AFSCs as prescribed by TOPCAP. Air Force promotion and enlisted force managers may incorrectly assume that equal selection opportunity during each cycle ultimately yields equal promotion opportunity over a career. A key point is that the Air Force did not develop WAPS and the policy of equal selection opportunity in tandem. As this monograph unfolds, it will become clear that there is a subtle disconnect between the two that would be mitigated by standardizing WAPS test scores.

There is also a concern about AFSC mix in the pool of future enlisted leaders. The Air Force fills about 400 strategic enlisted leadership and management positions with individuals who have performed well in one or more previous jobs as chiefs. Because chiefs face mandatory retirement at 30 years of service, logic dictates that individuals who are promoted to E9 with 20 or fewer years of service are better postured to obtain the experiences they need to fill future strategic chief vacancies. Figure 1.2 shows that some AFSCs have higher percentages of young chiefs. For example, in Comm-Computer Systems (3C), 12 percent had 20 or fewer years of service in September 2006, compared to Security Forces (3P) which had only 4 percent. Because the Air Force does not standardize test scores, the by-AFSC inven-

---

15 Figure 1.1 captures 88 percent of the top three inventory. The remaining 12 percent are in small AFSCs or tax AFSCs.

Figure 1.2
Percentage of Chief Master Sergeants Within an AFSC With 20 or Fewer Years of Service, September 2006

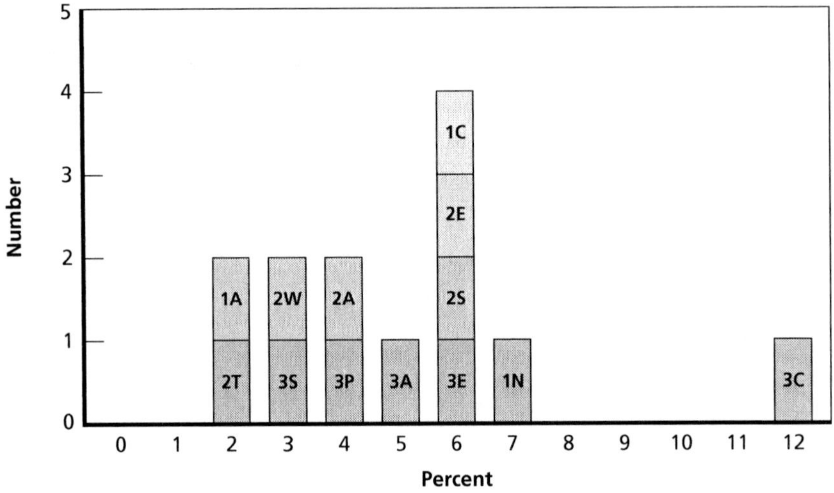

SOURCE: AFPC's Interactive Demographic Analysis System (IDEAS).
NOTE: The 13 largest non-tax, 2-digit AFSCs contained 81 percent (1,728/2,142) of the E9s from non-tax AFSCs and 64 percent (1,728/2,705) of all chiefs. These 13 2-digit AFSCs are actually made up of 89 5-digit AFSCs.
RAND *MG678-1.2*

tory of young chiefs varies randomly over time, which may not be consistent with strategic chief master sergeant requirements.

Finally, random variations in the influence of testing and longevity yield unpredictable results that must also be perplexing to many NCOs who believed they were well postured for promotion, which is inconsistent with the objective of a visible promotion system.

Organization of the Monograph

Chapter Two supplies background information on WAPS to include a discussion of the factors that actually drive selection results. Chapter Three provides background material on test score standardization techniques. It demonstrates that standardization is a common practice, and it discusses the reasons for standardizing some well-known tests—the

Scholastic Aptitude Test (SAT), the American College Test (ACT), and the Armed Services Vocational Aptitude Battery (ASVAB). Chapter Three also discusses various standardization strategies that would help achieve specific Air Force objectives. Chapter Four quantifies the relationships between testing impacts and selection timing, and it demonstrates the random and inconsistent nature of WAPS criteria. Using the foundation established in Chapters Two through Four, Chapter Five details four potentially adverse effects of not standardizing WAPS test scores. Two of these effects are internal promotion system issues. However, the other two at times have detrimental impacts on the strength management system. Finally, Chapter Six presents conclusions and recommends that the Air Force implement a standardization strategy that will produce predictable results that are compatible with its objectives.

CHAPTER TWO

The Weighted Airman Promotion System: Motivation, Mechanics, Reality, and Theory

WAPS is the system that the Air Force uses to promote airmen to the top five grades within each AFSC, and it is the instrument through which the Air Force strives to identify individuals who have the highest potential.

Shore and Gould provide insight into the motivation for WAPS:

> In the middle 1960s, the enlisted promotion system to E4 through E7 had command-centered promotion boards but no standard promotion procedures. Promotion eligibles had no understanding of how competitive they were and no one could give them guidance on how to improve their promotability if they were not promoted. Airmen dissatisfaction was growing and that dissatisfaction was being expressed to the air staff and congress in increasing volume. Congress was receiving similar mail from the other services' personnel but the volume from the Air Force was the greatest.
>
> During late 1967, congressional hearings on DoD enlisted promotions asked Major General Horace Wade (AF/DPX) penetrating questions about Air Force enlisted promotions, and he promised promotion system change. He then tasked the Air Force Human Resources Laboratory (AFHRL) with developing an objective and visible enlisted promotion system for E4, E5, E6, and E7s.

> AFHRL convened a panel to identify the relevant factors to consider, and then sit as a promotion board and rank eligibles from most to least promotable. "Policy capturing" methods were then used to mathematically capture the consensus policy of the board. This resulted in weights which were multiplied times each promotion factor and those products were summed to provide a total promotability score. Since that score ranked the candidates the same as the actual ranks assigned by the board, the board's policy had been mathematically captured. With those weights, additional eligibles could be ranked without the board members being present and those ranks would be the same that the board would have given had they been present.
>
> In July 1968, the Secretary of the Air Force approved a change from a board process for promoting E4 through E7 promotion eligibles to the weighted factor process if it could be proven that the system promoted the same airmen as an operational promotion board. This resulted in a 1969 field test of WAPS (the Weighted Airman Promotion System). The new PFE (promotion fitness exam) tests were taken to the Alaskan Air Command (AAC) and given to all E4 through E7 promotion eligibles. When the test scores were available WAPS scores were computed for all the eligibles and held in confidence. Meanwhile, centralized AAC promotion boards were convened and used current board procedures and full promotion folders to rank and select candidates for promotion. Then the AAC board rankings were compared to the WAPS rankings and the two systems consistently ranked and identified the same personnel for promotion.
>
> In 1970, the WAPS 6-factor system became operational . . . .[1]

Since 1970, the Air Force has promoted individuals to E5–E7 using a formula that does not involve inputs from selection boards.

---

[1] Shore and Gould, 2004, pp. 2–4.

Each eligible airman earns a weighted score that is a function of Enlisted Performance Report (EPR) scores, decorations, score on the annual Promotion Fitness Exam (PFE), score on the annual Specialty Knowledge Test (SKT), time in service (TIS), and time in grade (TIG).[2] Selection to E8 and E9 has an additional component—a board score. Each of these factors carries different point values that in 1970 reflected the importance that board members placed on that factor.

The Air Force revalidated WAPS in 1972, 1977, 1986, and 2004 (Appendix D). "Revalidated" is a bit misleading because each of these four efforts found that WAPS does not reflect what selection boards would reward for most grades. The fact that the Air Force did not subsequently alter the WAPS weighting factors suggests a strong aversion to change despite the system's prime directive to promote those with highest potential.

However, WAPS has changed in response to other pressures. Appendix C documents nine major changes to WAPS over time. It points out that each individual's percentile score originally determined his or her test points. In statistical terms, test points awarded for every AFSC had the same mean and standard deviation. In lay terms, this meant that testing had about the same impact in every AFSC. In 1972, the Air Force changed the method for determining test points to the percentage of correct responses. This change reduced the impacts of testing by reducing the ranges of test scores. As we show in this chapter, the change also meant that testing no longer had the same impact in every AFSC. As our analysis will indicate, these differential impacts yield consequences.

Table 2.1 lists the current WAPS weighting factors.

---

[2] Enlisted members take tests annually. Test scores from previous years have no impact on the current year's selection outcome.

Table 2.1
Current WAPS Factors

Factor	Purpose of Factor	Selection to:	Maximum Points
Specialty Knowledge Test (SKT)	E4s–E6s take an SKT annually to assess their knowledge about their specialty. For AFSCs without an SKT, and for SKT-exempt individuals, WAPS now doubles PFE scores.	E5–E7	100 points based on percentage of questions answered correctly.
Promotion Fitness Exam (PFE)	E4s–E6s take the PFE annually to assess their general knowledge about the Air Force.	E5–E7	100 points based on percentage of questions answered correctly.
Air Force Supervisory Exam	E7s and E8s take the Supervisory Exam annually to assess their general knowledge about the Air Force from management and leadership perspectives.	E8–E9	100 points based on percentage of questions answered correctly.
TIS	Rewards total years of Air Force experience.	E5–E7	Up to 40 points; 2 points for each year of total active military service up to 20 years—1/6 point per month.
TIS	Rewards total years of Air Force experience.	E8–E9	Up to 25 points; 1 point for each year of total active military service up to 25 years—1/12 point/month.
TIG	Rewards experience in the current grade.	E5–E9	Up to 60 points; 1/2 point for each month in grade up to 10 years
Decorations	Rewards outstanding performance that the Air Force has recognized with medals (decorations).	E5–E9	Up to 25 points; each decoration carries a point value of 0 to 15.

Table 2.1—continued

Factor	Purpose of Factor	Selection to:	Maximum Points
EPR scores	Supervisors provide written EPRs (normally, annually) to document performance. EPRs also contain an integer score of 1 to 5, with 5 being the best.	E5–E9	Up to 135 points; weight is given only to EPRs rendered in the past five years, with less weight on older EPRs and more weight on the most recent.
Board score	For each AFSC, a three-member board, composed of two chief master sergeants and a colonel who are subject matter experts, evaluates each member's record of performance to include the written portions of EPRs, duty history, and professional military education.	E8–E9	270–450 points

SOURCE: U.S. Air Force, 2005a, p. 272–273.

A Fundamental Relationship

One can surmise from Table 2.1 that WAPS allows individuals who are not strong testers[3] or who have few EPR or decoration points to compensate with increased longevity.[4] It also allows junior airmen who are extremely proficient testers to compete successfully for selection. For example, Table 2.2 illustrates that 3P0X1 (Security Forces) E5s who became E6s in FY06 had a broad range of 6–20 years of service. Implicitly, the Air Force deemed that all these individuals had approximately the same potential to serve as E6s.

[3] Hereafter, we refer to those who take tests as *testers*.

[4] In this analysis, we define *longevity* as the sum of TIS and TIG points.

Table 2.2
YOS Distribution for E5s in AFSC 3P0X1 Who Became E6s in FY06

YOS	Number Promoted
6	1
7	24
8	46
9	72
10	67
11	151
12	153
13	58
14	39
15	14
16	4
17	1
18	1
19	0
20	4

SOURCE: U.S. Air Force, 2006a.

Reality

To truly appreciate the driving forces in the enlisted promotion system, it is not sufficient to understand the mechanics of Table 2.1, which prescribes only how members earn points. WAPS factors with more points available do not necessarily play greater roles in determining who the Air Force promotes. Rather, the factors that have greater variations in points actually awarded have greater impacts on selection results. In other terminologies, variation measures range or point spread. For example, for selection to E7, EPR scores make up to 135 points available. However, because almost all E6s earn nearly perfect EPR scores, EPRs have the smallest point spread of the WAPS factors and the lowest impacts in determining selection to E7.[5]

[5] Within the Air Force personnel analysis community, it is fairly common knowledge that WAPS outcomes are driven by the variations in component scores. See Duncan, 1994.

Figure 2.1 illustrates that for AFSC 2A5X1 (Aerospace Maintenance) on the 05E7 cycle, about 85 percent of E6s earned perfect EPR scores.[6] Because EPR scores displayed little variation, perfect EPR scores made little difference in determining the rank order of those competing for selection to E7 (we show later that EPR scores play greater roles in selections to E5 and E6).

To reinforce the concept of variation, imagine an extreme case where every E6 in an AFSC had a perfect EPR score of 135. If we rank-ordered those E6s using just their longevity, testing, and decoration scores, subsequently adding 135 EPR points to every score would not alter the rank order.[7] In the case of AFSC 2A5X1 on the 05E7 cycle,

Figure 2.1
Distribution of EPR Scores on 05E7 for 2A5X1

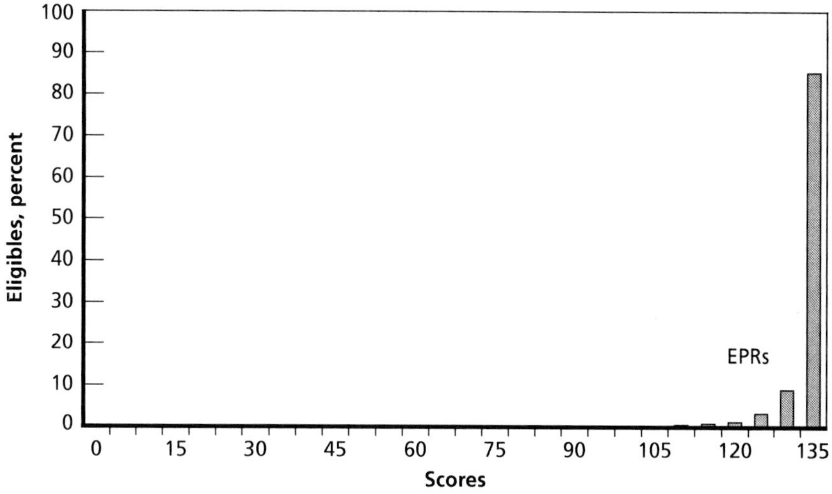

SOURCE: Derived from AFPC WAPS history file.
RAND *MG678-2.1*

[6] In the standard notation used to describe an Air Force selection cycle, the first two digits represent the year and the last two characters are the grade to which members competed for selection. There is occasionally a fifth character, "A" or "B," when there are two selection cycles to the same grade in the same year.

[7] We combine PFE and SKT scores to yield a testing score. PFE and SKT also have a positive correlation.

when we rank-order the 946 eligible E6s based only on decoration, testing, and longevity points, adding EPR points changes the average person's rank order by only 3.0 percentile points.

For AFSC 2A5X1 on the 05E7 cycle, decoration scores, with fewer points available, had more variation in points awarded than did EPR scores (Figure 2.2).

For AFSC 2A5X1 on the 05E7 cycle, when we rank-order the eligible E6s based only on EPR, testing, and longevity points, adding decoration points changes the average person's percentile by 5.2 points. Hence, by this method of estimating the impact of WAPS factors for this AFSC on this cycle, decorations would have almost twice the influence on rank order as EPRs.

Figure 2.3 illustrates even greater variation in longevity scores for AFSC 2A5X1 on the 05E7 cycle.

After we rank-order the eligible E6s based only on EPR, decoration, and testing points, adding longevity points changes the average person's percentile by 11.5 points.

Figure 2.2
Distribution of Decoration Scores on 05E7 for AFSC 2A5X1

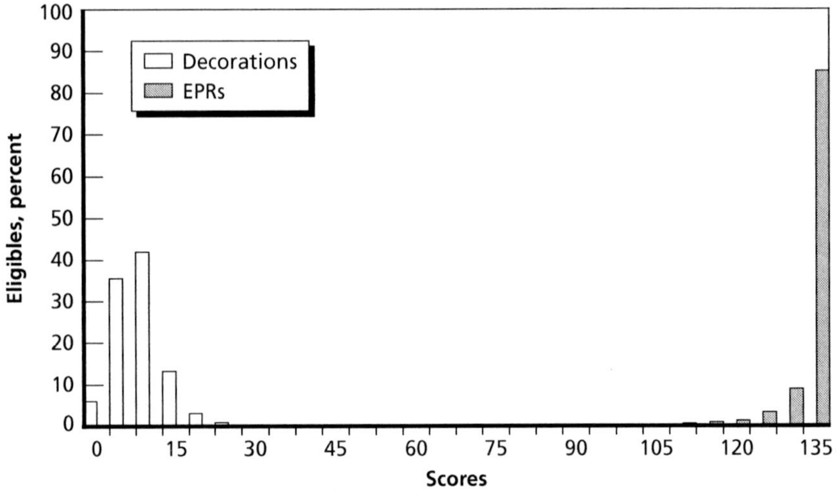

SOURCE: Derived from AFPC WAPS history file.
RAND *MG678-2.2*

Figure 2.3
Distribution of Longevity Scores on 05E7 for AFSC 2A5X1

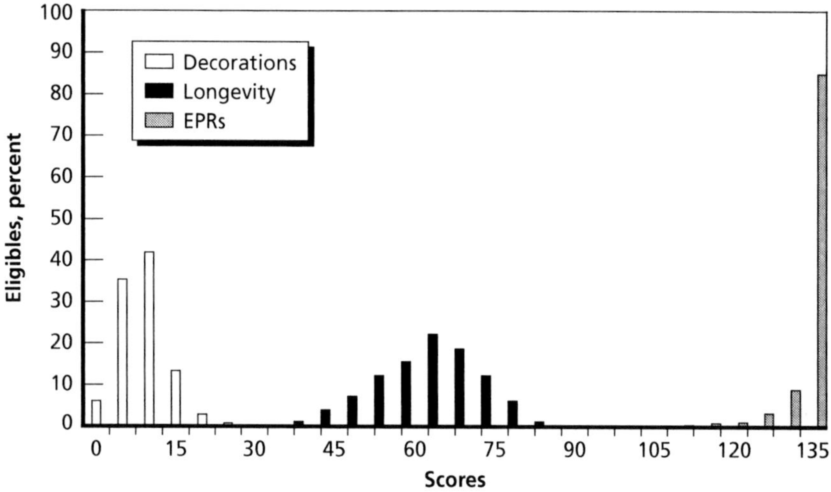

SOURCE: Derived from AFPC WAPS history file.
RAND *MG678-2.3*

The WAPS factor that displayed the greatest variation for AFSC 2A5X1 on the 05E7 cycle was testing (Figure 2.4). After we rank-order the eligible E6s based only on EPR, decoration, and longevity points, adding testing points changes the average person's percentile by 24.3 points, which is over twice the impact of longevity.

Because AFSCs comprise members with different mixes of general and specialty knowledge, because tests can vary in difficulty across AFSCs, and because the Air Force does not standardize test scores, test-score variation is not the same in every AFSC within a cycle. In addition, the variation in test scores is generally not the same within an AFSC over time. Figure 2.5 illustrates the difference in test score variations for two AFSCs for the 05E7 cycle. For AFSC 3E2X1 (Pavement and Construction Equipment), most of the test scores ranged between 80 and 160 points—a spread of 80. For AFSC 7S0X1 (Special Investigations), the spread was 70 (90 to 160), with

Figure 2.4
Distribution of Testing Scores on 05E7 for AFSC 2A5X1

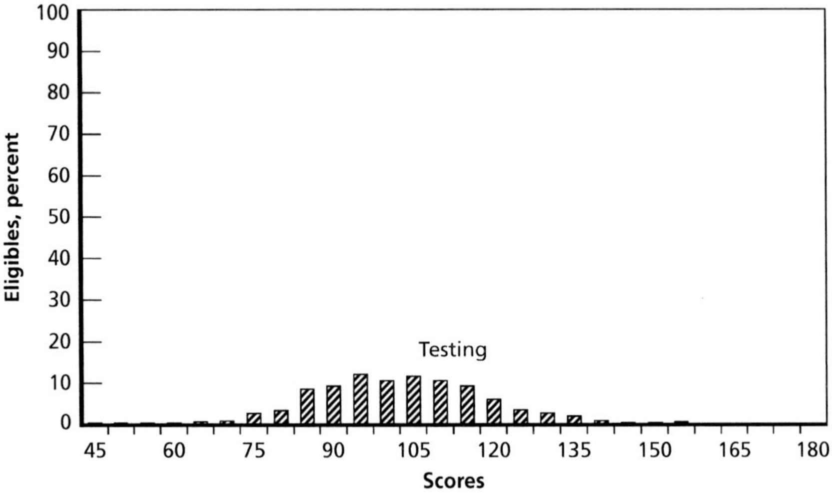

SOURCE: Derived from AFPC WAPS history file.
RAND *MG678-2.4*

much greater concentration near the center of that spread. Therefore, the scores for AFSC 7S0X1 were concentrated in a narrower range with reduced impacts on selection outcomes, all other things being equal.

These differences in the spread of testing scores could have been due in part to the basic abilities of the testers in the two AFSCs. Figure 2.6 shows that there were differences in the distribution of Armed Forces Qualification Test (AFQT) scores of the E6 testers in the two example AFSCs for this cycle.[8] However, the greater dispersion of basic abilities in AFSC 7S0X1 population would predict a corresponding greater dispersion of WAPS test scores in that AFSC than the observed greater concentration relative to AFSC 3E2X1. Hence, more-dominant factors must have been in play.

[8] Testing experts in the Office of the Secretary of Defense (OSD) derive the AFQT score from a subset of the ASVAB test modules that is primarily related to math and English skills. An AFQT percentile score represents an individual's ability relative to the general U.S. youth population.

Figure 2.5
Distribution of Testing Scores on 05E7 for AFSCs 3E2X1 and 7S0X1

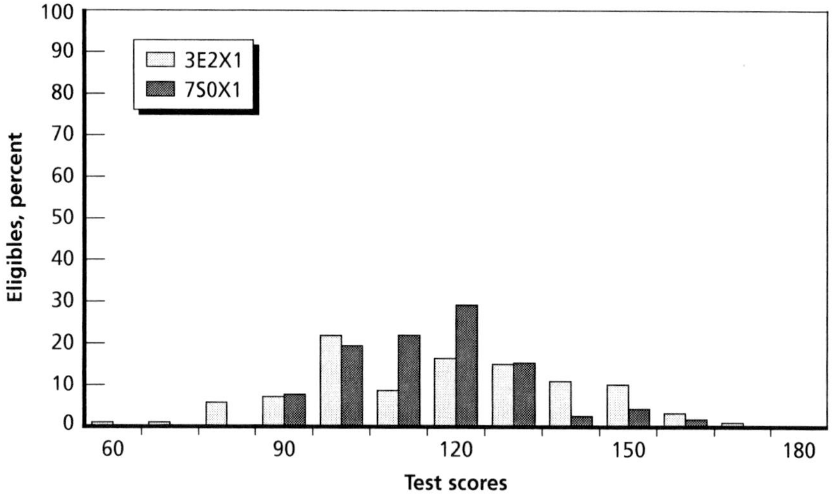

SOURCE: Derived from AFPC WAPS history file.
RAND *MG678-2.5*

Figure 2.6
Distribution of AFQT Scores of 05E7 Testers for AFSCs 3E2X1 and 7S0X1

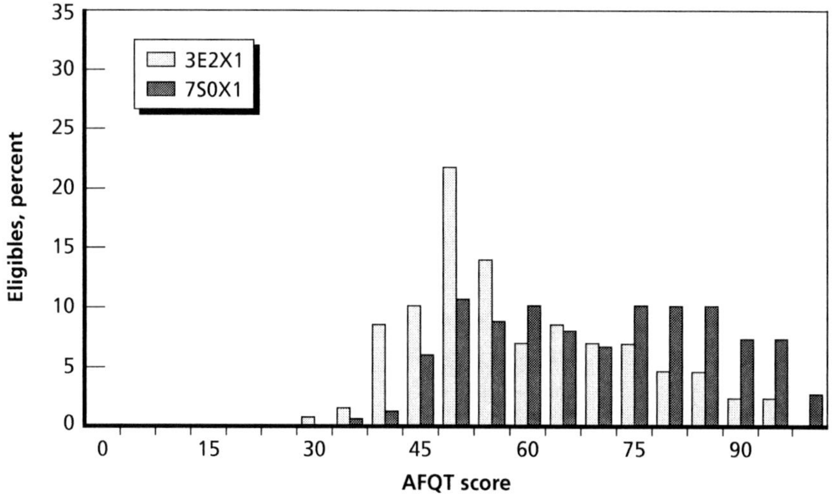

SOURCE: Derived from AFPC WAPS history file and the personnel data system.
RAND *MG678-2.6*

Theoretical Impacts of Differences in Variation

For selection to E7, the greatest variation in awarded WAPS points occurs in test scores. Partially because the Air Force does not standardize test scores across AFSCs, testing does not produce the same selection impacts in every AFSC. Some might argue that testing differences across AFSCs are not problematic because members of a given AFSC only compete among themselves for selection. However, this view misses the mark. Differences in testing impacts enable some AFSCs to produce higher top three/E6 ratios. In Chapter Four, we present detailed data to support the following serpentine logic trail:

1. Suppose an AFSC has a difficult SKT or a difficult version of the PFE relative to the quality of its testers. Tests that are more difficult (to a point) lead to wider ranges of test scores.[9] Within an AFSC, this tends to favor good testers by placing more weight (variation) on testing and less weight on longevity and the other WAPS factors.
2. Better testers tend to have less TIG. Members with good EPR scores and more TIG necessarily have histories of poor testing. Therefore, in AFSCs with difficult tests (and large variations in test scores), the good (younger) testers are better postured to overcome the fact that their more-senior competition—which is composed of poorer testers—have more longevity points. Hence, in these AFSCs, promoted airmen tend to have less time in service on average.
3. Consider an AFSC that has an E6 eligible base that is always 1,000, and suppose that the annual selection rate to E7 is 20 percent. Under ESO, that AFSC would realize 200 selections to E7 each year. Further suppose that the SKT for that AFSC is difficult relative to the quality of its testers. Because the range

[9] In an extremely simple test, all testers get 100 percent of the questions correct, and there is no variation in scores. In the most difficult test with multiple-choice questions, all testers randomly guess, and there is very little variation in test scores. Somewhere between these extremes, there are sets of questions that good testers can answer but poor testers cannot. These sets of questions yield the maximum variation in test scores.

of SKT scores would be large, testing would play a greater role in the selection outcome, and longevity would play a lesser role. In this case, suppose that the average phase point to E7 was 14 years.[10] These newly promoted E7s would compete on average for selection to E8 for 12 annual selection cycles before reaching high year of tenure at 26 years of service. Put another way, there would be 12 groups of 200 E7s in this AFSC (less losses to attrition and selection). Now, suppose the senior NCOs who develop the SKT for the same AFSC started writing much easier test questions. In subsequent administrations of the SKT, the range of test scores would decrease, testing would have less impact, and longevity would have a greater impact. Suppose that the impact was so great that the average phase point increased to 16 years of service. Newly promoted E7s would now compete on average for selection to E8 for 10 annual selection cycles before reaching high year of tenure at 26 years of service, and there would be 10 groups of 200 (less losses to attrition and selection) E7s in this AFSC.

Since ESO distributes selections to an AFSC based on the size of its eligible pool, it follows that when an AFSC has a lower phase point to E7, it is better postured to generate E8s and subsequently E9s (which also affects AFSC grade manning). This phenomenon may also exist at lower grades, but strength managers can compensate for its unfavorable manning impacts. However, strength managers have fewer options for dealing with E7, E8, and E9 manning deviations.

In practice, it is not easy to observe this theoretical phenomenon in the personnel data because the Air Force continually makes the following adjustments that affect the strengths of AFSCs:

- Retraining people to move into or out of an AFSC
- Career job reservations
- Selective reenlistment bonuses

---

10 *Average phase point* refers to the average years of service that a group has when it reaches the next grade.

- Higher selection rates for some chronically undermanned AFSCs
- Differential rates of drawing individuals from AFSCs for special duties
- The widespread practice of combining multiple AFSCs into a single AFSC at either E8 or E9
- Frequently merging or splitting AFSCs across all grades
- Using tests that are not consistently difficult for every grade within an AFSC
- Using tests that are not consistently difficult within grades for every AFSC.

We believe that the large number of options available to enlisted force managers and the inconsistent difficulty of tests explain why the issues we raise in this monograph have remained less visible.

Measuring Variation

Despite all the moving parts in the enlisted management system, we were able to confirm our theory with personnel data. Our first challenge was to develop an appropriate methodology to quantify the variation in WAPS components, and we considered four approaches (described in detail in Appendix E). Using the best of the approaches (Approach Four[11]), we show below that testing does not yield the same relative impacts in every AFSC. In turn, we use this information in Chapter Four to explain the selection implications for various groups based on seniority.

E5 WAPS Component Impacts

Figure 2.7 graphically illustrates the results of applying Approach Four to the 138 AFSCs that had at least 25 eligibles for the 05E5 cycle.

---

[11] For each AFSC, selection cycle, and WAPS factor, Approach Four calculates the average absolute difference in the number of standard deviations an individual is away from the mean, after including and excluding the points from each factor.

Figure 2.7
WAPS Component Impacts, 05E5 Cycle

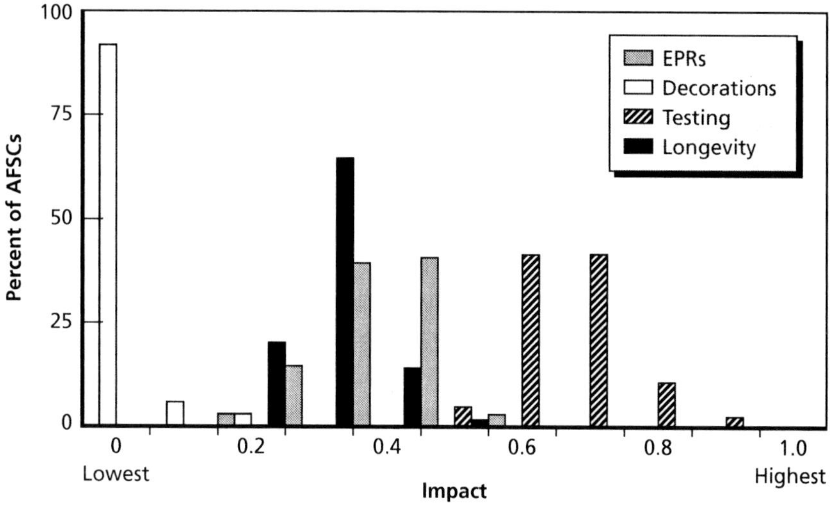

SOURCE: Derived from WAPS history file.
RAND *MG678-2.7*

For E4s competing to E5, testing had the greatest impact, followed by EPRs. For example, the hachured bars indicate that in 41 percent of the AFSCs, testing had a relative impact of 0.6 (rounded). For two percent of the AFSCs, testing had a relative impact of 0.9 (rounded). Figure 2.7 also illustrates that testing did not have the same impact for every AFSC. For AFSCs in which testing had the highest impact, either there was a greater range of test scores or other WAPS components had smaller-than-average variations.

Figure 2.8 shows that within some AFSCs, testing did not have consistent impacts over time. For these and other AFSCs, testing impacts spanned both high and low extremes over the eight E5 cycles we examined.

In Figure 2.7, the gray bars show that EPRs had the second-highest impact. For the 05E5 cycle, only 55 percent of E4s had perfect EPR scores. Because EPR impacts ranged from 0.2 to 0.6, we can deduce that members in some AFSCs were more likely to have

Figure 2.8
Within-AFSC Testing Impacts, E5 Cycles

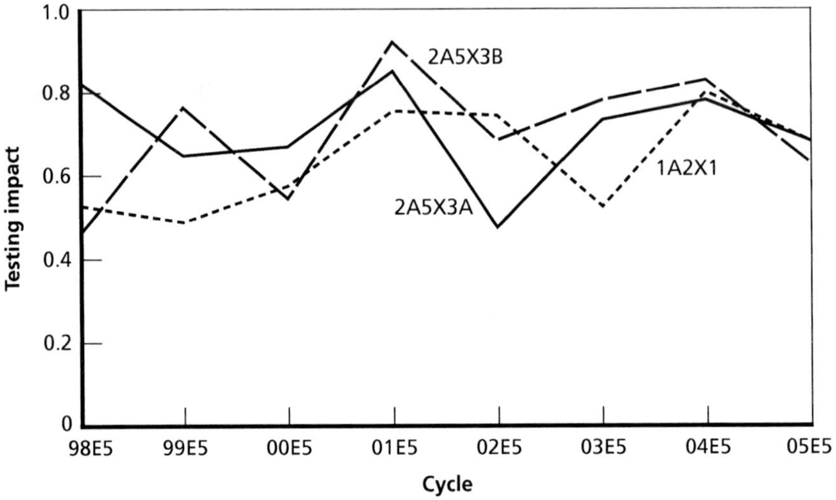

SOURCE: Derived from WAPS history file.
RAND *MG678-2.8*

perfect EPR scores than those in others. We confirm this in Figure 2.9, which illustrates that for the 05E5 cycle, 80 percent of testers in 10 AFSCs had perfect EPRs. At the other extreme, only 20 percent of the members in one AFSC had perfect EPR scores.

The black bars in Figure 2.7 indicate that longevity had slightly less impact than EPRs. Because longevity did not have the same impact in every AFSC (ranging from 0.2 to 0.5), we can suspect that the members of every AFSC did not possess the same TIS/TIG distributions. This could have occurred either because of differences in the percentage of members who initially enlisted for six years of service, because of disaggregate accession fluctuations, or because of retraining actions.

The white bars in Figure 2.7 show that decorations had the smallest impact (0.0–0.2). This was because most E4s had about the same number of decoration points (E4s are junior and have not had many opportunities to earn decorations).

To reinforce the preceding discussion, Figure 2.10 plots testing impacts, as measured in Approach Four, as a function of the standard

Figure 2.9
Different Rates of Perfect EPR Awards
(138 AFSCs in Cycle 05E5 with at least 25 eligibles)

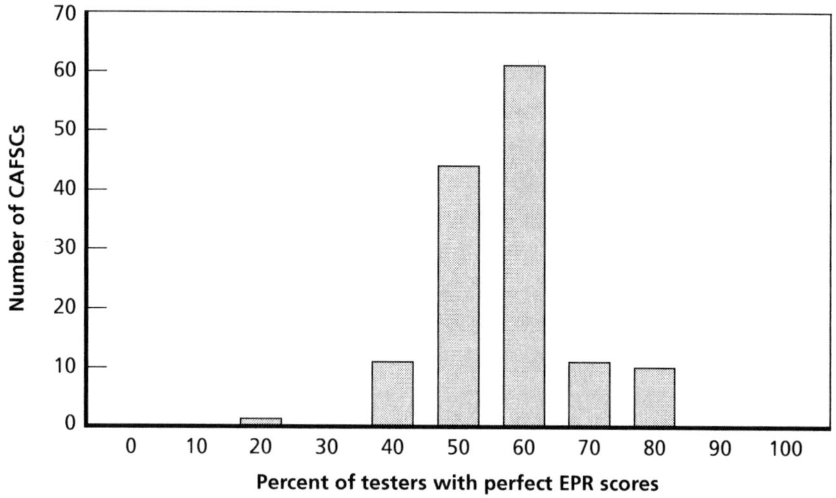

SOURCE: Derived from WAPS history file.
RAND *MG678-2.9*

Figure 2.10
Testing Impact Versus Standard Deviation of Test Scores

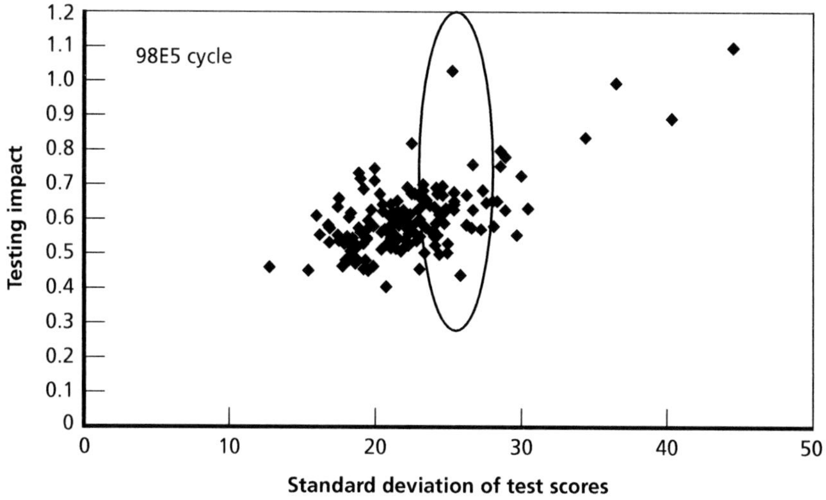

SOURCE: Derived from WAPS history file.
RAND *MG678-2.10*

deviations in test scores for the 98E5 cycle (each diamond represents an AFSC with at least 25 eligibles). The figure demonstrates that in general, greater variations in test scores translated into higher testing impact. However, as we observe for the AFSCs captured by the oval, there was a substantial range of testing impacts for AFSCs with nearly the same standard deviations in test scores.[12]

The vertical variation within the oval could occur only if there were differential variations in at least one of the other WAPS factors. To demonstrate that this was the case, Figure 2.11 plots testing impact versus the ratio of the standard deviation of test scores to the standard deviation of the total WAPS score for each of the 98E5 AFSCs.

Figure 2.11
Testing Impact Versus Standard Deviation in Test Scores Divided by Standard Deviation in Total Score, Cycle 98E5

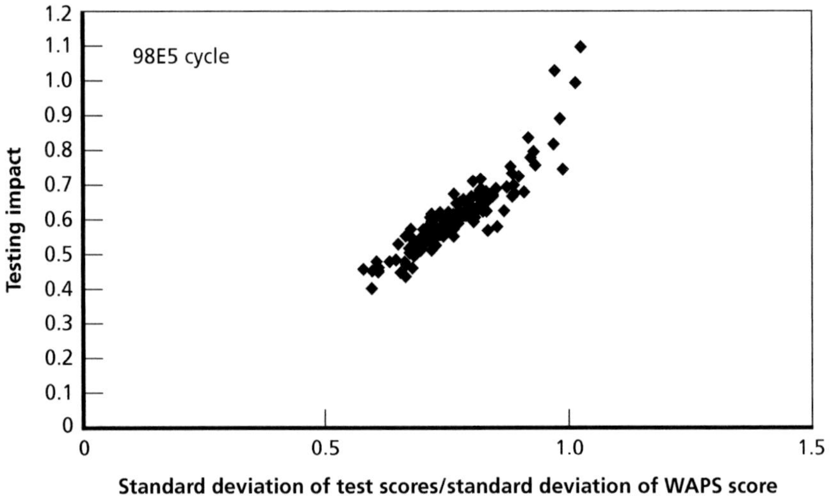

SOURCE: Derived from WAPS history file.
RAND *MG678-2.11*

[12] Even for AFSCs not in the oval, at best knowing the standard deviation of test scores would allow us to predict testing impact for an AFSC within only a 0.3 range.

The figure indicates that an AFSC's testing impact is closely related to its test score variation as a percentage of total WAPS score variation.[13]

For the 05E5 cycle, the relative impacts of WAPS factors as measured by Approach Four actually correspond to available WAPS points. As we demonstrate below, this relationship does not hold for other grades.

E6 WAPS Component Impacts

Figure 2.12 shows the distributions of impacts that we derived using Approach Four for the 143 AFSCs with at least 25 eligibles in the 05E6 cycle. Testing had the highest impact—and the range of impacts was about the same for 05E6 cycles as it was for 05E5 cycles.

Figure 2.13 shows that, within some AFSCs, testing was inconsistent and displayed a wide range of impacts over time.

Figure 2.12
WAPS Component Impacts, 05E6 Cycle

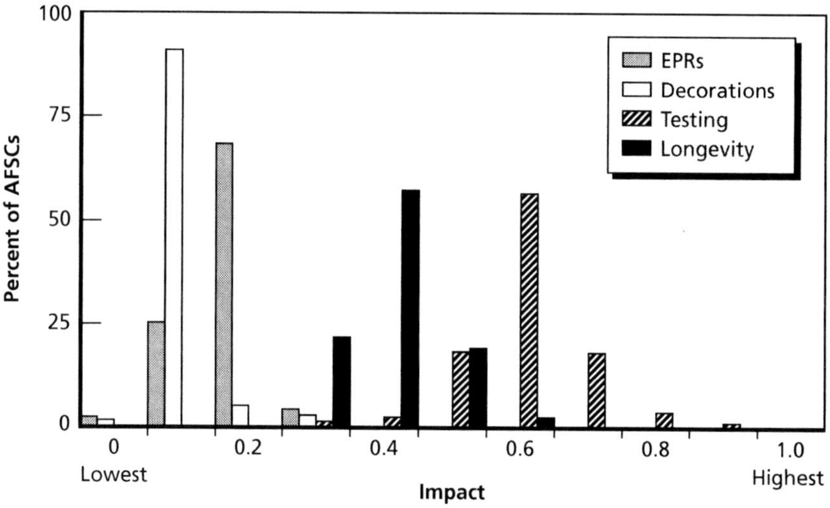

SOURCE: Derived from WAPS history file.
RAND *MG678-2.12*

[13] Knowing both the standard deviation of an AFSC's test scores and the standard deviation of its total WAPS scores, we could now predict testing impact within a 0.2 range, even for the outliers in Figure 2.10.

Figure 2.13
Within-AFSC Testing Impacts, E6 Cycles

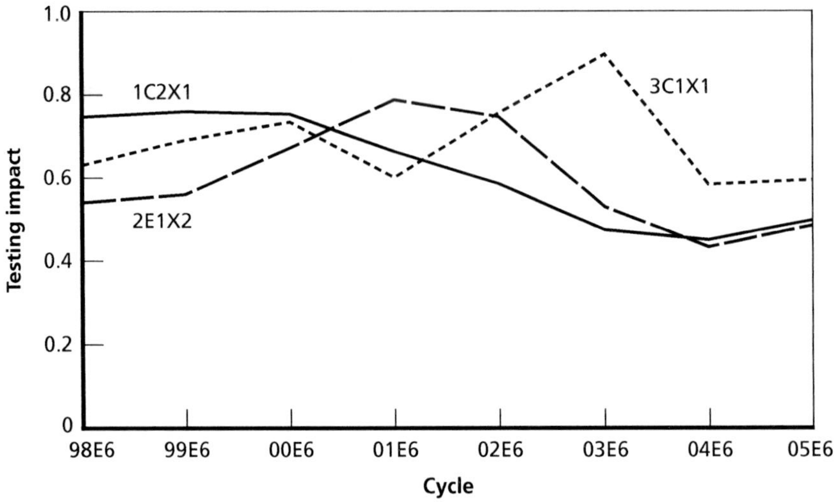

SOURCE: Derived from WAPS history file.
RAND *MG678-2.13*

A comparison of Figure 2.7 and 2.12 also indicates that longevity had much greater impact for the 05E6 cycle than it did for the 05E5 cycle. This occurred because E5s tended to have over twice the range of years of service as E4s (see Figure 2.14).

Compared to the E5 cycle, EPRs had less impact on the E6 cycle because E5s were more likely than E4s to have perfect EPR scores. For the 05E6 cycle, 67 percent of E5s had perfect EPR scores (compared with 55 percent of E4s).[14] Finally, as we would expect, decorations had a slightly greater impact for E5s because there was a larger variation in longevity (and hence, in the opportunity to earn decorations). From an aspiring fast burner's[15] perspective, even though a higher percentage of

---

[14] In addition, as we noted in Table 2.1, less weight is placed on older EPRs earned by E4s.

[15] The term *fast burners*, in Air Force parlance, refers to those promoted well ahead of their peers with the same length of service.

Figure 2.14
Distribution of Years of Service, September 2005

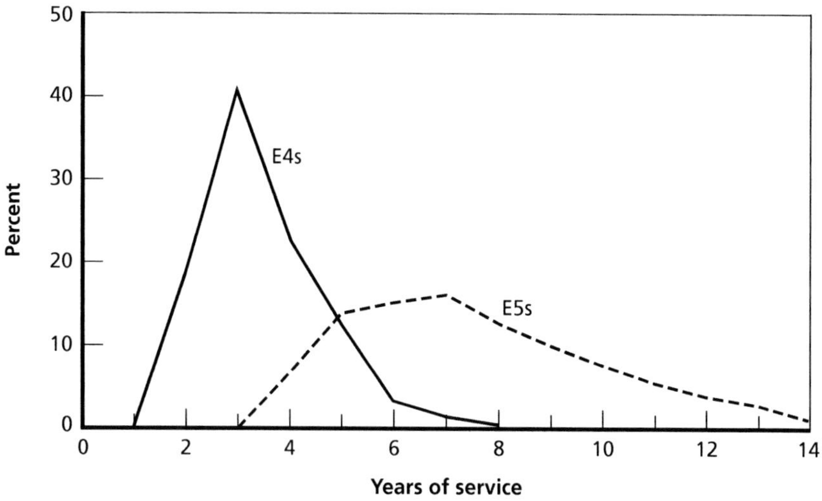

SOURCE: AFPC's Interactive Demographic Analysis System.
RAND *MG678-2.14*

E5 competitors had more longevity points, it was still possible to outpace them with strong test scores.[16]

E7 WAPS Component Impacts

Figure 2.15 shows the distributions of the statistics derived using Approach Four for 143 AFSCs with at least 25 eligibles in the 05E7 cycle. EPRs had reduced impacts because 84 percent of E6s had perfect EPR scores. Longevity also had reduced impacts, in part because TIG points are capped at 60 (10 years TIG) and TIS points are capped at 40 (20 years TIS). Testing had the highest impact for the 05E7 cycle, in part due to the reduced variations in EPR and longevity scores.

[16] Chapter Four shows that the ability to overcome fewer longevity points with strong testing is AFSC-dependent. Good testers who find themselves in AFSCs where it is more difficult to overcome less longevity have a reduced opportunity for promotion.

Figure 2.15
WAPS Component Impacts, 05E7 Cycle

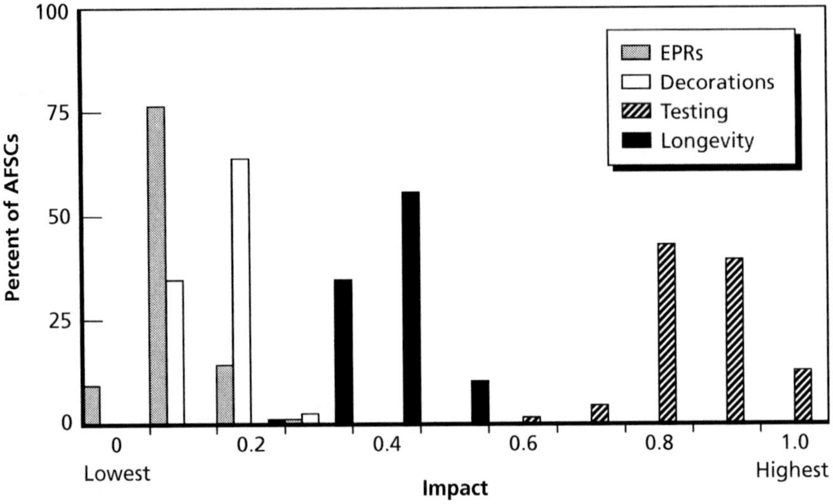

SOURCE: Derived from WAPS history file.
RAND *MG678-2.15*

Figures 2.16–2.18 illustrate why testing had the highest impact for the 05E7 cycle. They plot the distributions of the standard deviations of test scores and total scores for the 05E5–05E7 cycles and demonstrate the growth of testing variation as percentages of the total WAPS score variation.

Figures 2.16 and 2.17 indicate similar relationships between the distributions of test score standard deviations and total score standard deviations for the 05E5 and 05E6 cycles. Consistent with these relationships, we saw in Figures 2.7 and 2.12 that testing had approximately the same impact.[17] However, in Figure 2.18, we see that testing accounted for almost all the variation in total score standard deviations. This implies that there was a relatively small range in the standard deviations of the other WAPS components. Consequently, testing

---

[17] Although we do not display the data, the relationships in Figures 2.16–2.18 held for all the 98–05 cycles.

Figure 2.16
Distribution of Standard Deviations of Test and Total Scores, 05E5 Cycle

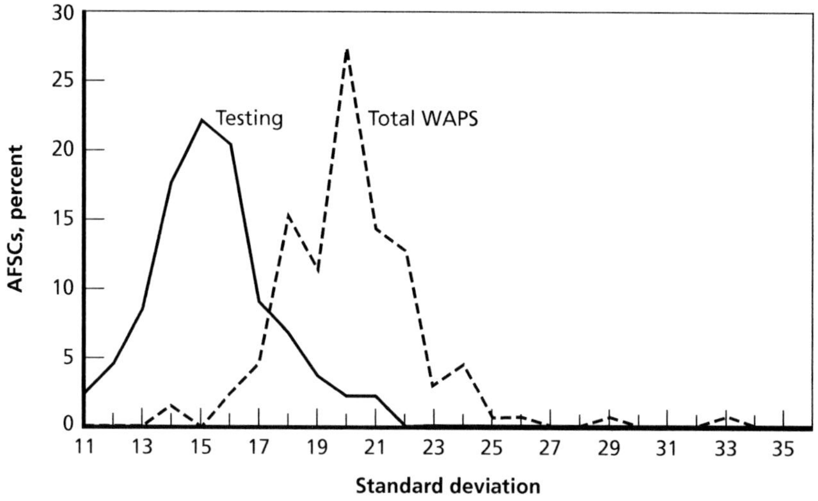

SOURCE: Derived from WAPS history file.
RAND *MG678-2.16*

Figure 2.17
Distribution of Standard Deviations of Test and Total Scores, 05E6 Cycle

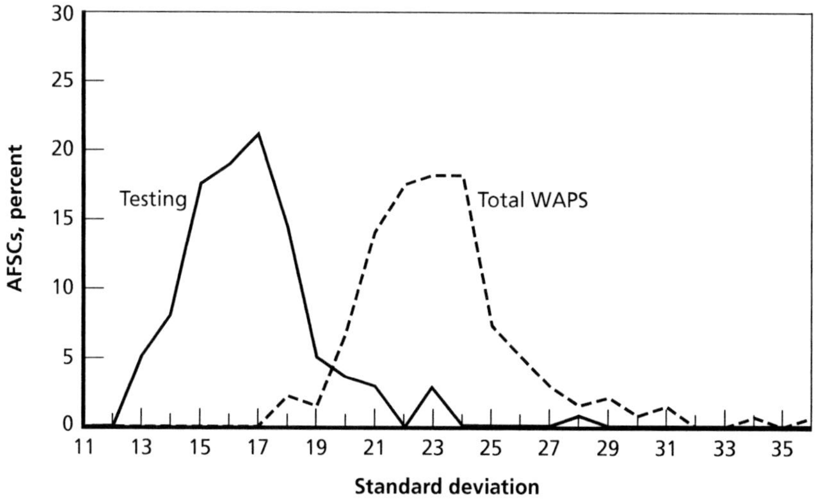

SOURCE: Derived from WAPS history file.
RAND *MG678-2.17*

Figure 2.18
Distribution of Standard Deviations of Test and Total Scores, 05E7 Cycle

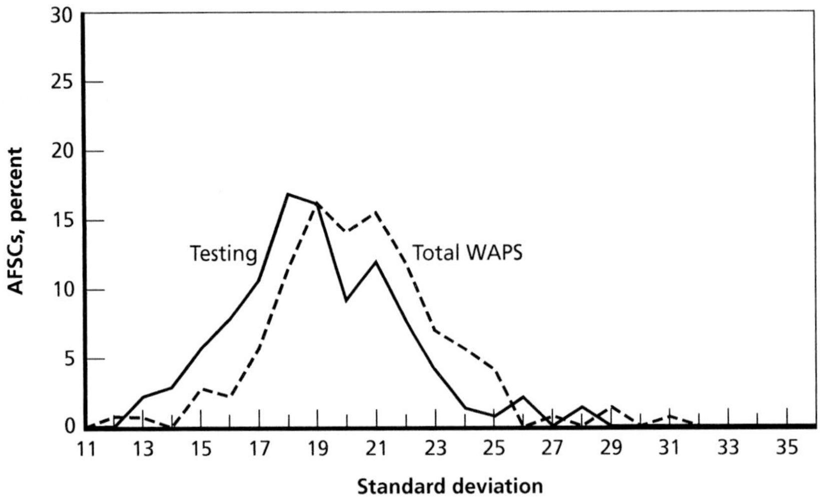

SOURCE: Derived from WAPS history file.
RAND *MG678-2.18*

had a much greater impact in the E7 cycles. We also observe in Figure 2.18 that the distribution of the standard deviations for test scores had a broader range than for the 05E5 and 05E6 cycles.

Finally, Figure 2.19 shows that, within some AFSCs, testing impacts varied widely over time.

E8 WAPS Component Impacts

There are three major differences in E8 and E9 promotions compared to E5–E7 promotions. First, SNCOs competing for selection to E8 and E9 meet selection boards that award between 270 and 450 points. Second, the Air Force Supervisory Exam, which reduces available testing points to 100, replaces the PFE and SKT. Finally, the maximum number of TIS points drops to 25. As we would expect, the ranges of testing and

Figure 2.19
Within-AFSC Testing Impact, E7 Cycles

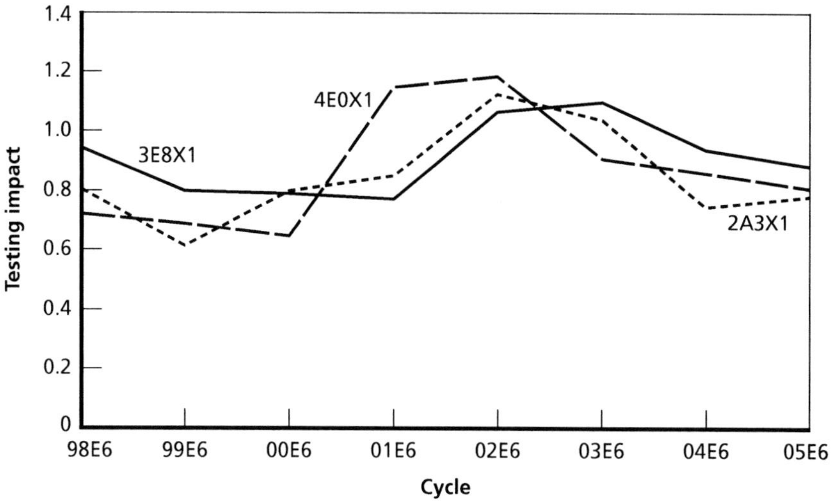

RAND *MG678-2.19*

longevity points diminish and both play lesser roles in selections to E8 and E9. In Chapter Four, we discuss some of the factors that influence E8 and E9 board scores.

Figure 2.20 shows the distributions of WAPS component impacts that we derived using Approach Four for the 99 AFSCs with at least 25 eligibles on the 05E8 cycle. Generally, board scores had the greatest impact, although the impact was not the same for every AFSC.[18]

E9 WAPS Component Impacts

Figure 2.21 shows the distributions of statistics derived using Approach Four for the 26 AFSCs with at least 25 eligibles for the 05E9 cycle. As with the E8 cycle, board scores dominated. Hence, we can deduce

---

[18] Members of each AFSC meet different boards that are composed of chief master sergeants and colonels who are subject matter experts. Figure 2.20 suggests that some boards awarded a greater range of scores than did others and hence had a greater impact on the selection outcomes.

Figure 2.20
WAPS Component Impacts, 05E8 Cycle

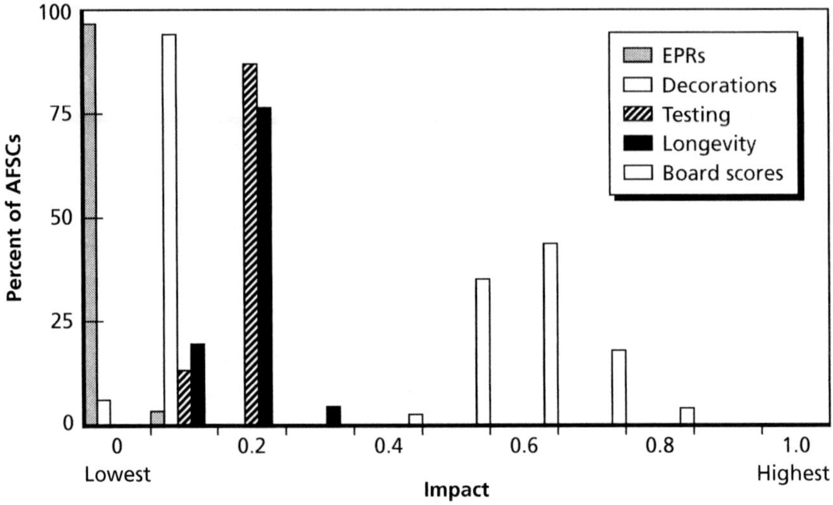

SOURCE: Derived from WAPS history file.
RAND *MG678-2.20*

Figure 2.21
WAPS Component Impacts, 05E9 Cycle

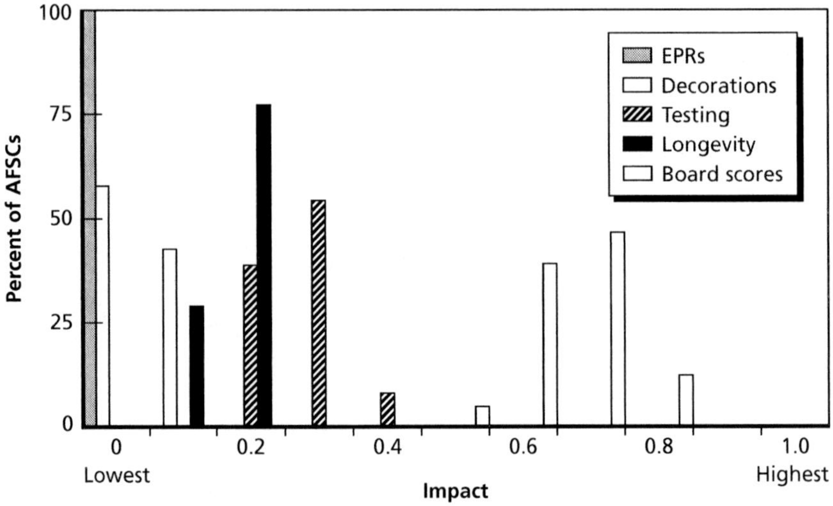

SOURCE: Derived from WAPS history file.
RAND *MG678-2.21*

from Figures 2.20 and 2.21 that standardizing scores on the Supervisory Exam would have little impact on E8 and E9 selection outcomes (which we later verify in Figure 5.6).

Chapter Summary

Thus far, we have established that WAPS factors do not have the same impact on selections to different grades. In addition, they do not have the same impacts on AFSCs within grades or within AFSCs over time. We show in Chapters Three and Four that these differences are not by design. Hence, the random inconsistencies in promotion impacts make it difficult to assert that people with the highest potential are consistently selected, as specified in AFPD 36-25. In Chapter Five, we discuss additional implications of these differences. Our objective there is to demonstrate that WAPS may be yielding unrecognized and undesirable results.

CHAPTER THREE

Standardizing Test Scores

Ability and knowledge testing to qualify for advancement opportunities is not unique to the Air Force. For example, many if not most high school seniors nationwide take one or more tests to qualify for such opportunities as admission into colleges or universities or entrance into the U.S. armed forces. Each enlisted member of the Air Force has become qualified to join the Air Force in part by taking and scoring well on the Armed Services Vocational Aptitude Battery (ASVAB). Ability and knowledge testing does not end with high school. Many occupations require testing to qualify for promotions within organizations. For example, the City of Los Angeles requires passing ability and knowledge tests to qualify for higher-level jobs.[1]

In general, we can categorize tests into two groups—those that are standardized and those that are not. The Air Force uses nonstandardized raw scores from different forms and administrations of PFEs and SKTs to determine selection outcomes. In this context, a raw score is the percentage of questions answered correctly regardless of the form or administration of the test. By comparison, well-known tests taken for college admission purposes and even the ASVAB are standardized tests. In these tests, the converted raw score enables meaningful comparisons of different forms and administrations of the test.

This chapter and Appendix I examine the standardization procedures of common tests that have properties and purposes similar to PFEs and SKTs. We selected the American College Test (ACT) developed

[1] Telephone communication with J. Kawai, Los Angeles Fire Department, January, 2006; Los Angeles, 2006.

by ACT, Inc., the SAT I Reasoning Test developed by the Educational Testing Service, and the Armed Services Vocational Aptitude Battery developed by the Defense Manpower Data Center of the Department of Defense. These tests, like PFEs and SKTs, are primarily multiple-choice, and all evaluate test takers for advancement opportunities.

What Is Test Standardization?

In the strictest sense, a test is termed standardized if the same score represents the same level of knowledge and/or ability across multiple applications and versions or forms of the test. Hence, standardization can have implications for the design of the test as well as for the method by which the performance on the test is reported. A standardized test is designed according to specific design formats. A nonstandardized test need not abide by stringent design rules, but often does abide by similar design regularity constraints. A standardized test does not report performance by the raw score on the test. Rather, the raw score is transformed into a measure of performance through mathematical processes that make the measure of performance comparable to those of different administrations and versions of the test. A nonstandardized test simply uses the number or percentage of correct answers as its performance measure and no attempt is made to link any aspect of the test to any other test or version of the test that is used for the same purpose.

Why Standardize?

Standardization makes the scores on different forms or versions of a test directly comparable. Having comparable measures means that opportunities can be made equally available to all those who demonstrated comparable levels of ability.[2] Therefore, equity of opportunity is the ultimate reason for standardizing test scores.

---

2 Performance on a test is often only one of many criteria that determine whether an opportunity for advancement is actually afforded to candidates. Hence, standardization of test

Approaches to Standardizing PFE/SKT Scores

The Air Force has avoided the costs associated with standardizing PFEs and SKTs because it does not have a pressing requirement to compare WAPS test scores over time. Because of concerns about test compromise, the Air Force does administer multiple versions of the PFE.[3] However, because it does not standardize scores, the Air Force requires all members of an AFSC to take the same version of the PFE. Hence, the Air Force can only use multiple versions to deter compromise across AFSCs, not within AFSCs.

If the Air Force desired to compare scores across test administrations, it would need to embark on a complex path of test score standardization. Case studies that involve applying various equating methods to historical data would be one approach to begin the process. For example, to identify the specific method that would be most applicable to the Air Force selection tests, the primary characteristics of the Air Force tests would need to be considered. Trials using different equating approaches, in case study form, would reveal the adequacies of the various standardization approaches. The Air Force would also need to understand the potential racial and gender implications of test score standardization. A transition plan that addressed the potential impacts of a different scoring method would also need to be developed. Such a plan would have to consider the psychological and perceived sensitivities to changing from the familiar and straightforward raw score method now in use to a method that provides less visibility but is potentially fairer.

However, because the Air Force does not need to compare test scores over time, it would not be necessary to employ complex and costly standardization techniques (detailed in Appendix I). This chapter discusses a less elaborate approach to standardization that would suffice to address the concerns we raise in this monograph.

---

scores helps ensure all those of the same ability will qualify for advancements but is most often not used as the sole determinant of which candidates are actually advanced.

3 We understand from our conversations with test developers at the Occupational Measurement Squadron that the Air Force plans to move back to a single version of the PFE.

Standardization Mechanics

The Air Force could standardize the raw scores of the individuals within an AFSC using the following simple equation:

$$x_s = [(x_0 - \mu_0) \times (\sigma_s / \sigma_0)] + \mu_s \tag{3.1}$$

where

x_s = standardized score
x_0 = raw score
μ_0 = mean (average) of raw scores in the AFSC
σ_s = desired standard deviation of standardized scores
σ_0 = standard deviation of raw scores in the AFSC
μ_s = desired mean (average) of standardized scores.

When using Equation 3.1, it is not necessary to simultaneously change both the mean and standard deviation of a set of raw scores. For example, Figure 3.1 shows the distribution of the raw SKT scores for AFSCs 3E2X1 (Pavement and Construction Equipment, an AFSC with high testing impact) and 7S0X1 (Special Investigations, an AFSC with low testing impact) for the 05E7 cycle. The mean (μ) and standard deviation (σ) for AFSC 3E2X1 were 60.04 and 14.26 and, for AFSC 7S0X1, 49.35 and 9.33.

If the Air Force wanted the standard deviation for each AFSC's SKT scores to be 11.00 without changing mean scores, it would simply apply Equation 3.1 and set $\mu_s = \mu_0$ for each AFSC. Figure 3.2 plots the results of those transformations for our example.

Currently, raw scores do not communicate to individuals how well they performed relative to their peers. An individual could score 45 on the SKT one year and 55 the following year, with the 55 actually representing a poorer performance relative to the competition. To make scores more meaningful, the Air Force could also use Equation 3.1 to standardize means. Figure 3.3 shows the results of standardizing both the means and standard deviations of our example AFSCs. In this case, we arbitrarily set the standardized means to 50 and the standard deviations to 11. These choices mean that about 95 percent of the indi-

Figure 3.1
Distributions of Raw SKT Scores for AFSC 3E2X1 and 7S0X1, Cycle 05E7

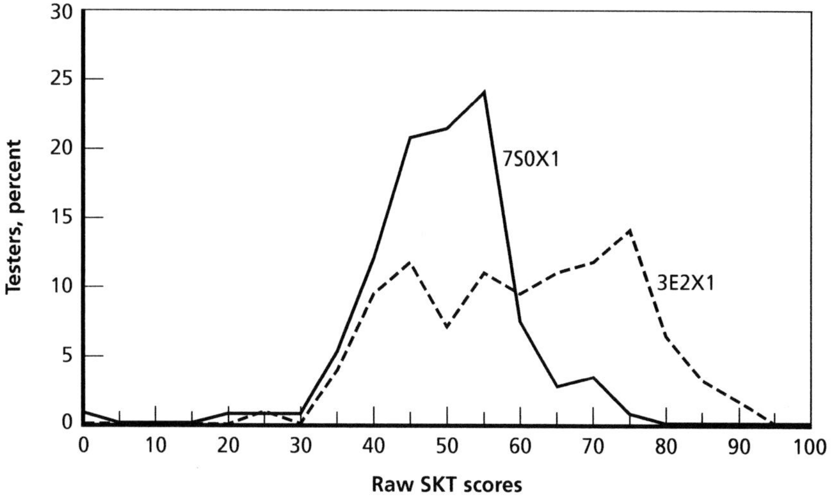

SOURCE: WAPS history file.
RAND *MG678-3.1*

Figure 3.2
Distributions of Standardized SKT Scores for AFSCs 3E2X1 and 7S0X1, σ_s =11, Cycle 05E7

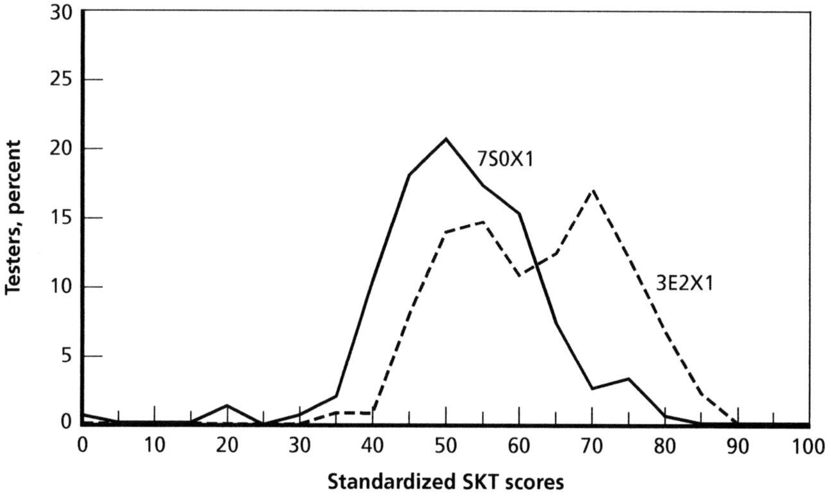

SOURCE: Derived from WAPS history file.
RAND *MG678-3.2*

Figure 3.3
Distributions of Standardized SKT Scores for AFSCs 3E2X1 and 7S0X1, $\mu_s = 50$, $\sigma_s = 11$, Cycle 05E7

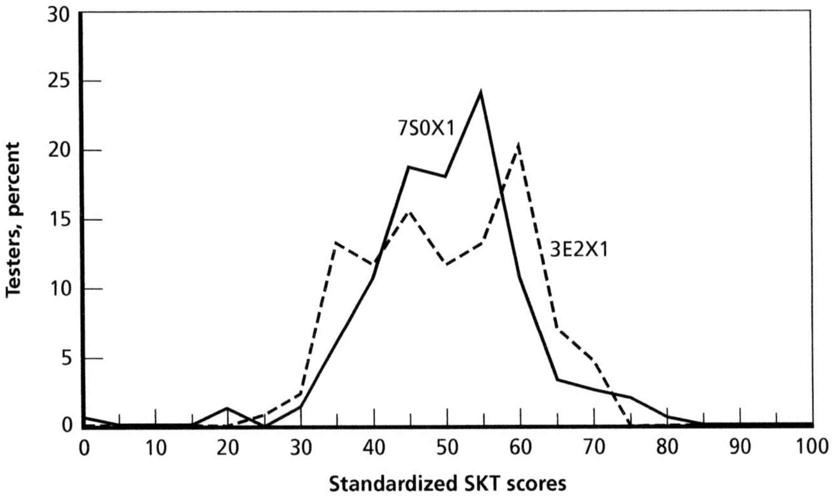

SOURCE: Derived from WAPS history file.
RAND *MG678-3.3*

viduals would have standardized SKT scores in the range (mean ± two standard deviations) or (50 ± 22) or (28 – 72). With a standardized mean of 50, an individual who scored a standardized score of 55 would know that his/her score was slightly above average. (Enlisted members currently receive feedback about their testing performances relative to the competition through their Promotion Score Notices and published averages for selects and non-selects.)

An Alternative Approach to Standardization

If the Air Force is not open to mechanically standardizing scores, the experts who develop the PFEs and SKTs at the Air Force Occupational Measurement Squadron indicate that, by exercising more control during the test development phase, they could influence the means and standard deviations of test scores.[4] This would offer an indirect approach to standardization. However, such an approach would be more expensive and less precise than applying Equation 3.1.

Disclaimer

Using Equation 3.1 would only allow the Air Force to convey to individuals through a standardized score where they stood relative to the competition tested in a particular selection cycle. However, this simple approach would not measure trends in absolute abilities over time.

---

[4] Currently, the Occupation Measurement Squadron periodically assembles subject matter experts from across the Air Force to develop questions for the SKTs. There is little control over whether or not these experts develop easy or difficult tests. Hence, testing impacts vary randomly depending upon the inclinations of the test development teams.

CHAPTER FOUR

Testing Impact and Selection Timing

In Chapter Two, we established that WAPS factors do not yield the same impacts on selections to different grades; do not have the same impacts on AFSCs within grades; and do not produce the same impacts within AFSCs over time. In this chapter, we discuss the relationships between testing impacts and selection timing. This will lay the groundwork for Chapter Five, which explains the linkage between selection timing and potentially undesirable results.

Selections to E2–E4

In the enlisted force, selection to E2 through E4 is on a fully qualified basis and is primarily a function of time (Table 4.1). The Air Force promotes 15 percent early to E4 based on performance.[1] In addition to meeting these longevity milestones, members must satisfy performance requirements. There are also a number of avenues for accelerated selections.[2]

Fixed phase points to E4 provide financial predictability to young airmen who are making car payments or planning to start fami-

[1] Commanders also delay promotions for a small percentage of airmen with disciplinary or low-performance issues.

[2] Under certain conditions, the Air Force grants accelerated promotion in some AFSCs, for ex-service academy cadets, for college semester hours completed, for high school ROTC, for Civil Air Patrol achievements, for scouting achievements, and so forth.

Table 4.1
Typical Phase Points to E2–E4

	Four-Year Enlistee	Six-Year Enlistee
To E2	6 months of service	Not applicable
To E3	16 months of service	Completion of basic military training, usually at six weeks of service
To E4	36 months of service	29.5 months of service

SOURCE: U.S. Air Force, 2002.

lies. They also eliminate the requirement to develop and administer selection tests for a large segment of the enlisted force.[3]

Selections to E5

Many factors influence an individual's opportunity for selection to E5. We demonstrate below that the impact of testing in an AFSC is one of those factors. Because we cannot graphically isolate the true impacts of testing, we also discuss our regression models that relate selection rates for junior through senior E4s to multiple factors.[4] Our objective

[3] Fixed phase points to E4, under current accession policy, mean that force managers cannot simultaneously achieve 100 percent 5-level and E5 manning. This statement may cause some to take notice, but the explanation is straightforward. First, over 90 percent of 5-levels are either E4s or E5s. This means that if force managers are to satisfy 5-level manning requirements, they must be able to control the sum of the E4 and E5 inventory. Force managers can, and do, control the number of E5s by promoting enough E4s to fill E5 vacancies. However, force managers cannot control the number of E4s. E4 losses are driven in large part by selections to E5. However, the E4 grade is unique because its losses cannot be replaced by promoting E3s—because phase points are fixed to E4. Therefore, force managers are unable to control E4 gains, which are functions of fluctuating accessions in previous years, first-term attrition, and the mix of 4- and 6-year of service accessions. Because force managers cannot control the number of E4s, they cannot control 5-level manning (and simultaneously keep aggregate E5 manning at 100 percent). It is interesting to note that when WAPS was originated in 1970, selection to E4 was part of WAPS, which permitted force managers to better control skill-level manning by varying E4 selection rates. However, force managers lost that flexibility in 1971 when the Air Force fixed phase points to E4.

[4] We initially developed models to predict phase points. However, these models had little power because phase points, particularly to higher grades, are influenced by past force man-

is to demonstrate that, all other things being equal, the Air Force is more likely to promote junior E4s from AFSCs in which testing has greater impacts than their contemporaries from AFSCs with lower testing impacts. If this is the case, the Air Force is not applying the same standards to identify the most-qualified personnel in each AFSC.

Figure 4.1 shows the extraordinary range of selection rates that individuals with four years TIS realized in each AFSC in the 05E5 cycle. At one extreme, two AFSCs realized selection rates of 18 percent (rounded). At the other extreme, two AFSCs had selection rates of 48 percent. This differential emphasis on experience further highlights the random fluctuations within the enlisted promotion system. In this chapter, we show that the majority of this gap is due to differences in testing impacts, which the Air Force could control by standardizing test scores.

Figure 4.1
Selection Rates, Four Years TIS, Cycle 05E5

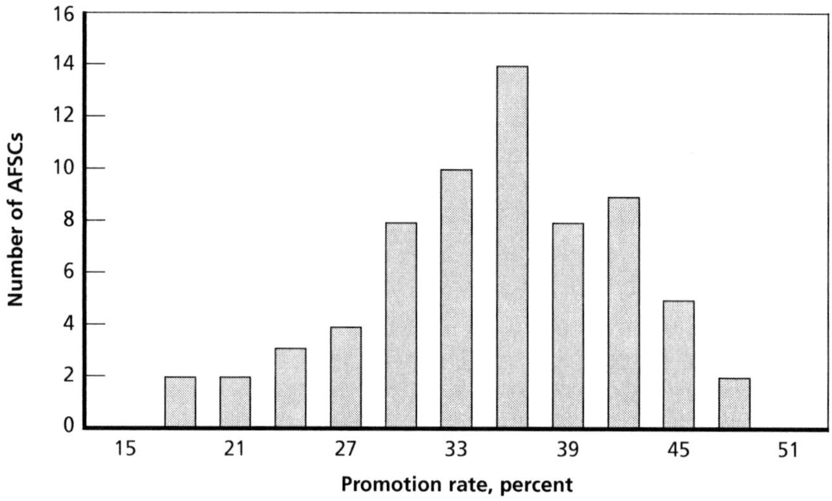

SOURCE: WAPS history file.
RAND *MG678-4.1*

---

agement decisions and randomly changing WAPS component impacts. By focusing on single cycle selection rates, we eliminated the daunting prospect of sorting out force management and WAPS impacts and interactions over the past 30 years.

To understand the reasons for the range of selection rates in Figure 4.1, we first developed a graphical approach.

A Univariate Perspective of Selections to E5

Our objective in this section is to demonstrate visually that testing does not yield the same timing impact for every AFSC in E5 selection cycles. In Appendix F we demonstrate this phenomenon mathematically, but that discussion may be less intuitive to those who are not familiar with linear regression.

In this and the corresponding analyses for selection to grades E6 and E7, we did not use every AFSC. First, we eliminated all special-duty AFSCs because the individuals in those AFSCs are nonhomogenous cross-flows from many different AFSCs. We also eliminated small AFSCs that had fewer than 25 eligibles in a selection cycle.[5] After these adjustments, 132 AFSCs remained in our analysis of the 05E5 cycle.

Figure 4.2 plots the 05E5 selection rates for those with four through seven years TIS (rounded, as of the last month of the selection cycle) as a function of each AFSC's testing impact (using Approach Four, Chapter Two). To dampen fluctuations, we first ordered the AFSCs by increasing testing impact and then calculated moving averages. Each point in Figure 4.2 is a moving average that represents the pooled selection rate for all the eligibles in 20 AFSCs.[6] For example, the extreme left point for those with four years TIS indicates that the 05E5 pooled selection rate was 30.3 percent.[7] The second point from the left removes individuals from one AFSC and adds those from another to yield a moving average selection rate of 29.7 percent.

Figure 4.2 indicates that for E4s with four years TIS, the selection rates in AFSCs in which testing had the highest impact were about

---

5 There is nothing profound about our choice of 25. Because we subsequently consider the number of eligibles by time in service, including small AFSCs would yield many AFSC/TIS combinations with no eligibles.

6 We did not calculate the average of 20 selection rates because many small AFSCs had a selection rate of 0 (e.g., none selected out of two eligible with TIS = 4). Averaging rates would have given disproportional weight to small AFSCs.

7 We did not include a scale for the horizontal axis in Figure 4.2 because it represents an ordinal ranking.

Figure 4.2
Selection Rate Versus Testing Impact, 05E5 Cycle, 20 AFSC Moving Average

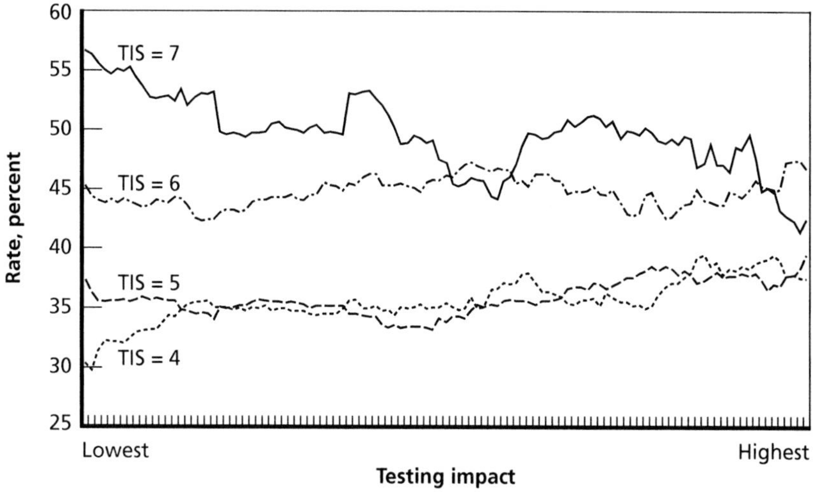

SOURCE: Derived from WAPS history file.
RAND *MG678-4.2*

eight percentage points higher (38 percent) than for E4s from AFSCs in which testing had the lowest impact (30 percent). By restricting the population to the junior E4s competing for E5, Figure 4.2 tends to compare E4s with very similar WAPS longevity points.

The group with seven years TIS is not as homogenous as the group with four years TIS because it is disproportionately missing previously promoted good testers from the AFSCs with high testing impacts. In the 20 AFSCs for which testing carried the lowest impact, those in the seven years TIS group had a selection rate of 57 percent.[8] These individuals, with the help of about 20 additional longevity points, were able to outpace the junior E4s in their AFSCs by 27 percentage points. In sharp contrast, in AFSCs with the highest testing impact, the promotion rate for E4s with seven years TIS was only five percentage points higher than for those with four years TIS.

---

[8] For each AFSC within a grade/cycle combination, testing impact is the same for members with greater and lesser TIS.

To complete the picture, Figure 4.2 also includes the groups with five and six years TIS, which fared as we would anticipate. Figure 4.3 displays the selection rates by TIS for the E4s in the extreme left and right groups of 20 AFSCs in Figure 4.2. For the 20 AFSCs in which testing had the highest impact, additional longevity points produced modest effects when compared with the AFSCs in which testing had the lowest impact.

A Multivariate Perspective of Selections to E5

To this point, we pooled E4s from 20 AFSCs and used moving averages to develop a sense for the impact of differential testing variations on an E5 cycle. Had we plotted selection rates for individual AFSCs instead of using moving averages, the TIS = 4 curve in Figure 4.2 would have looked like Figure 4.4.

The fluctuations in Figure 4.4 indicate that factors in addition to testing impact influenced 05E5 selection rates. To determine the true impact of testing, we needed to account for those factors.

Figure 4.3
Selection Rate Versus Highest and Lowest Testing Impact, 05E5 Cycle

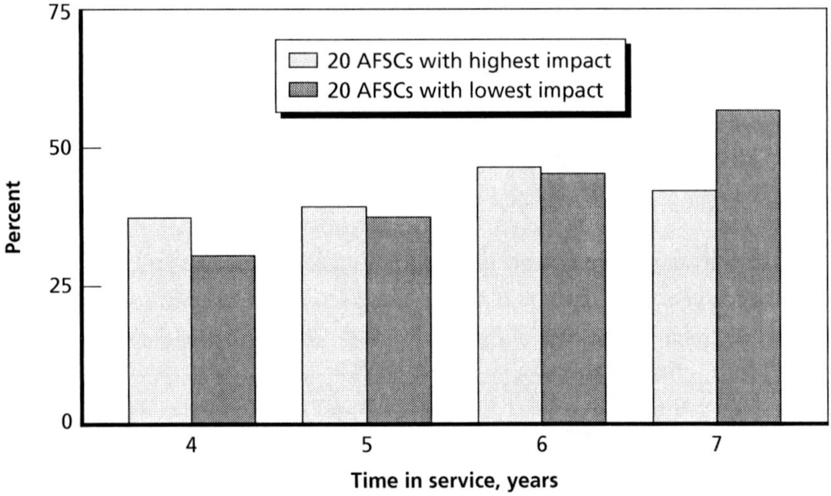

SOURCE: Derived from WAPS history file.
RAND *MG678-4.3*

Figure 4.4
Selection Rate Versus Testing Impact, TIS=4, 05E5 Cycle

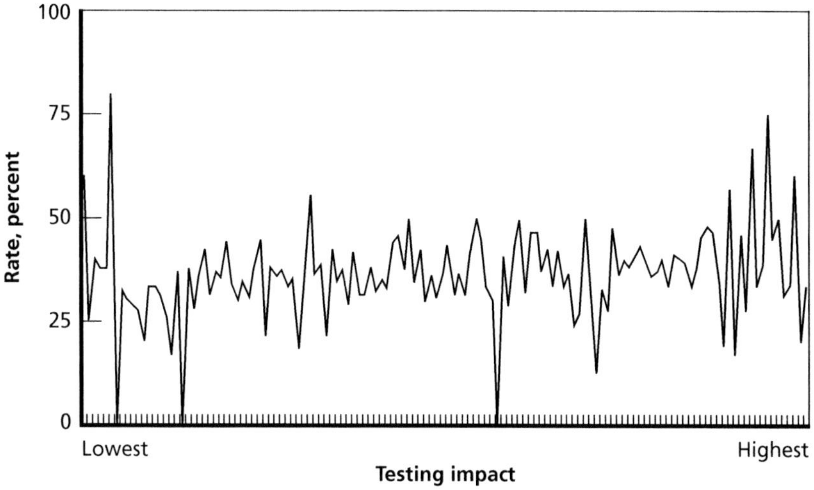

SOURCE: Derived from AFPC WAPS history file.
RAND *MG678-4.4*

Because so many variables might potentially influence enlisted selections, we could not isolate the impacts of testing without the aid of mathematical models. In Appendix F, we discuss four multivariate linear regression models that relate selection rates (our dependent or response variable) for junior through senior E4s (those with four, five, six, and seven years TIS) to these variables.

Figure 4.5 plots modeled selection rates, all other things being equal, for E4s in AFSCs with the highest and lowest testing impacts. It uses average values for EPR, decorations, and longevity impacts for non-CCS AFSCs.[9]

Figure 4.5 is revealing. It indicates that testing did not play a consistent role across AFSCs when the Air Force selected its most qualified personnel (except for those with six years of service). For AFSCs with the lowest testing impact, it also shows that modeled selection rates

[9] CCS stands for chronic critical shortage.

Figure 4.5
Selection Rate Versus Highest and Lowest Testing Impact, 05E5 Cycle

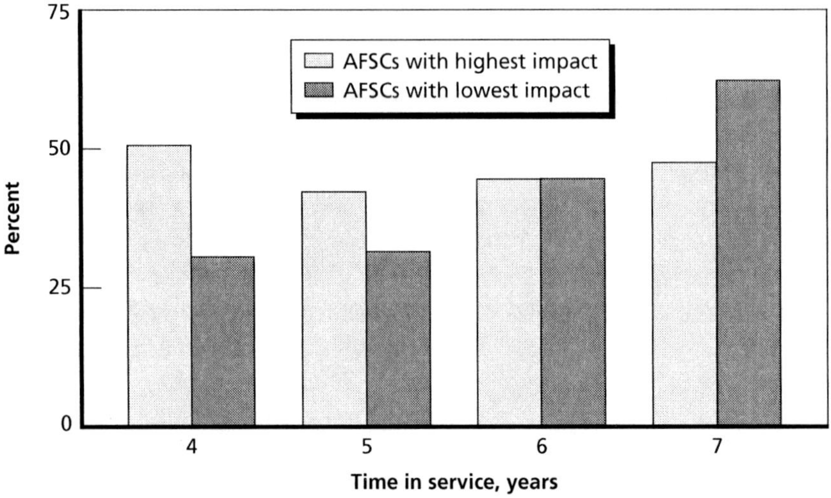

RAND *MG678-4.5*

increased with increasing longevity and that the Air Force promoted senior E4s at twice the rate of junior E4s. Conversely, for those in AFSCs with the highest testing impacts, additional longevity points were insufficient to allow senior E4s to compete successfully with junior E4s.

Selections to E6

As with selections to E5, multiple factors influence an individual's opportunity for selection to E6. In this section, we parallel our E5 analysis. Again, our objective is to demonstrate that, all other things being equal, the Air Force promotes junior E5s from AFSCs in which testing has a greater impact at higher rates than it does their contemporaries in AFSCs with lower testing impacts. If true, this would mean that the Air Force cannot be achieving its primary promotion objective.

The horizontal axis in Figure 4.6 shows the range of selection rates (rounded to the nearest 3 percent) that fast burners with seven or

Figure 4.6
05E6 Selection Rates, AFSCs with at Least 25 Eligibles with TIS ≤7

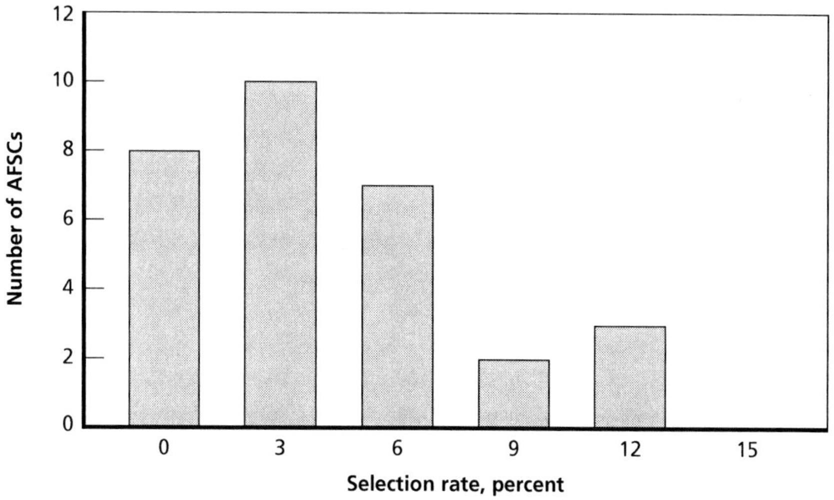

SOURCE: Derived from AFCP WAPS history file.
RAND *MG678-4.6*

fewer years TIS realized in each AFSC in the 05E6 cycle. At one extreme, eight AFSCs saw selection rates of 0 percent (rounded). At the other extreme, the Air Force promoted fast burners in three AFSCs at a 12 percent rate.

A Univariate Perspective of Selections to E6

Figure 4.7 plots the 05E6 selection rates for those with eight, ten, twelve, and fourteen years TIS as a function of testing impact. To dampen fluctuations, each point in Figure 4.7 is a moving average that represents the pooled selection rate for all the eligibles in 20 AFSCs.

Figure 4.7 indicates that for E5s with eight years TIS, the Air Force promoted those from AFSCs in which testing had the highest impact at almost four times the rate (13.3 percent) as E5s from AFSCs in which testing had the lowest impact (3.6 percent). For the poorer testers with 14 years TIS, the Air Force promoted individuals in AFSCs

Figure 4.7
Selection Rate Versus Testing Impact, 8, 10, 12, and 14 Years TIS, 05E6 Cycle, 20-AFSC Moving Average

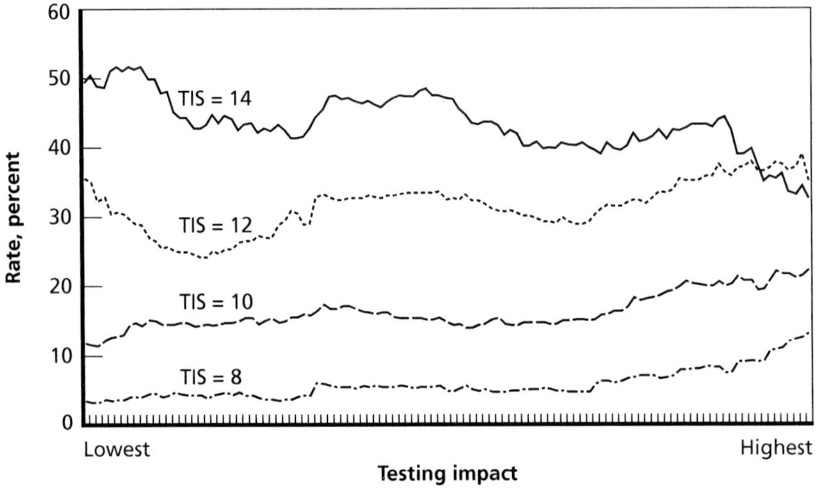

SOURCE: Derived from WAPS history file.
RAND *MG678-4.7*

that had the highest testing impact about 20 percentage points lower than those in AFSCs with the lowest testing impact.

Figure 4.8 captures the selection rates by TIS for the extreme left and right groups in Figure 4.7. For E5s with less than 12 years TIS, the Air Force promoted those in AFSCs with the highest testing impact about 10 points higher than their contemporaries. Beyond 12 years TIS, the Air Force promoted those in AFSCs with the lowest testing impact at higher rates.

A Multivariate Perspective of Selections to E6

Had we not used moving averages, the TIS = 8 curve in Figure 4.7 would have looked like Figure 4.9. The fluctuations in Figure 4.9 indicate that factors in addition to variations in testing influenced 05E6 selection rates. As we did in our E5 analysis, we developed models to help us isolate and quantify the true impact of variations in test scores. We developed five models to predict selection to E6 as a function of

Figure 4.8
Selection Rate Versus Highest and Lowest Testing Impact, 05E6 Cycle

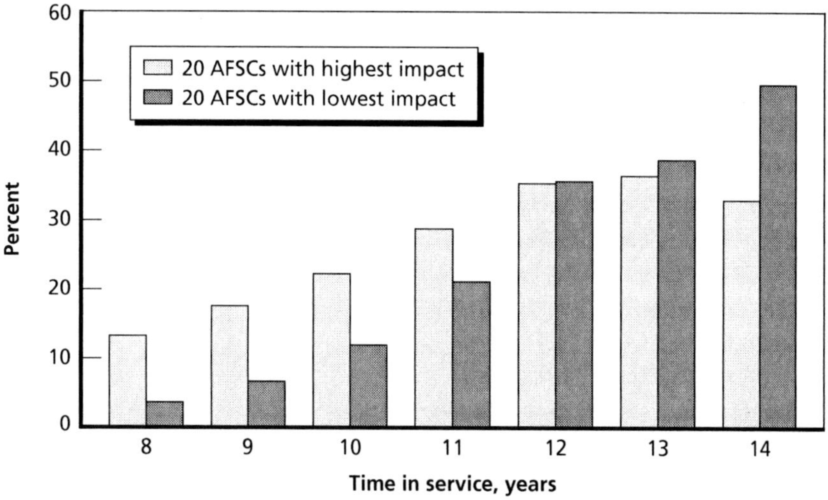

SOURCE: Derived from AFPC WAPS history file.
RAND *MG678-4.8*

Figure 4.9
Selection Rate Versus Testing Impact, 05E6 Cycle

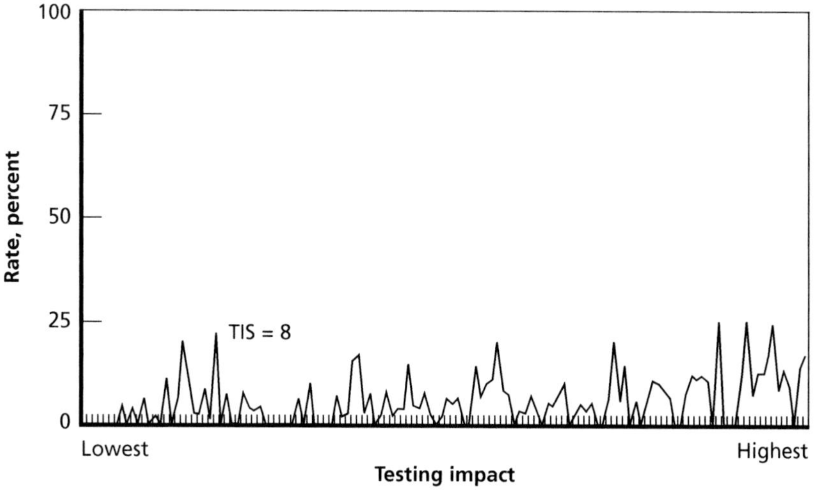

SOURCE: Derived from AFPC WAPS history file.
RAND *MG678-4.9*

TIS—fast burners (FB), those who were one or two years more senior than fast burners (FB+1+2), etc. See Appendix G for details. Figure 4.10 plots modeled selection rates to E6 in AFSCs with the highest and lowest testing impacts, all other things being equal. It uses average values for EPR, decorations, and longevity impacts for non-CCS AFSCs.

Figure 4.10 is telling. For AFSCs with the lowest testing impact, it indicates that selection rates increased with increasing longevity and that selection rates for senior E5s were almost 50 points higher than for junior E5s. In sharp contrast, for those in AFSCs with the highest testing impact, additional longevity points resulted in only a 20-point advantage for senior E5s.

Selections to E7

There are also multiple factors that influence an individual's opportunity for selection to E7. In this section, we parallel our E5 and E6

Figure 4.10
Selection Rate Versus High and Low Testing Impact, 05E6 Cycle

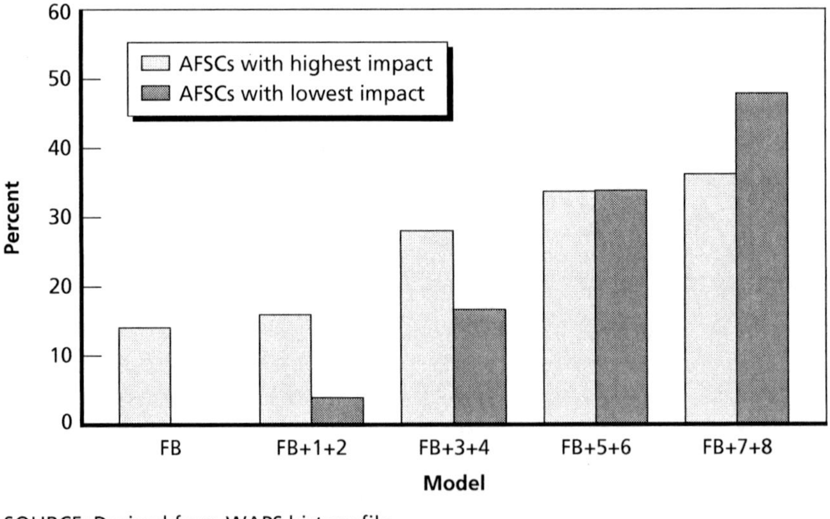

SOURCE: Derived from WAPS history file.
RAND *MG678-4.10*

analyses. Again, our objective is to demonstrate that, all other things being equal, the Air Force is more likely to promote junior E6s from AFSCs in which testing has greater impacts than their contemporaries in AFSCs with lower testing impacts.

Figure 4.11 shows the range of selection rates that individuals with 14 or fewer years TIS realized in each AFSC in the 05E7 cycle. At one extreme, the E6s in four AFSCs realized selection rates of 15 percent (rounded). At the other extreme, technical sergeants in one AFSC had a selection rate of 33 percent.

A Univariate Perspective of Selections to E7

Figure 4.12 plots the 05E7 moving average selection rates for three groups of E6s as a function of testing impact.

Figure 4.11
05E7 Selection Rates, AFSCs with at Least 25 Eligibles with TIS Less Than or Equal to 14

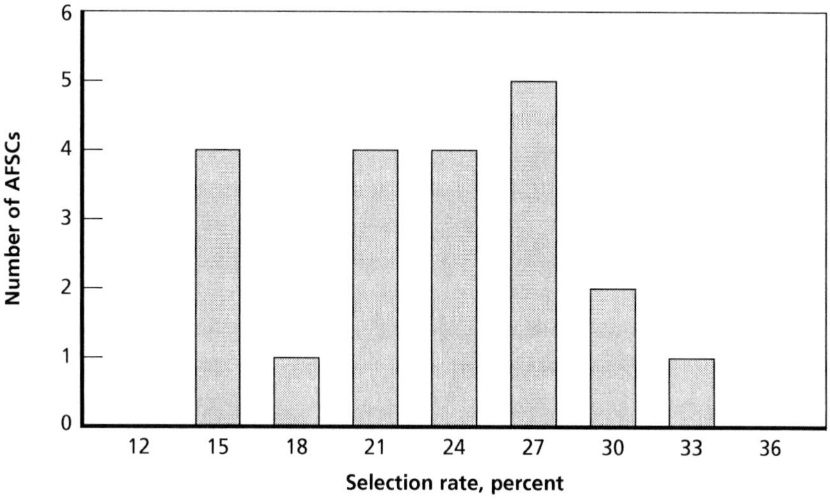

SOURCE: WAPS history file.

RAND *MG678-4.11*

Figure 4.12
05E7 Selection Rate Versus Testing Impact, 20-AFSC Moving Average

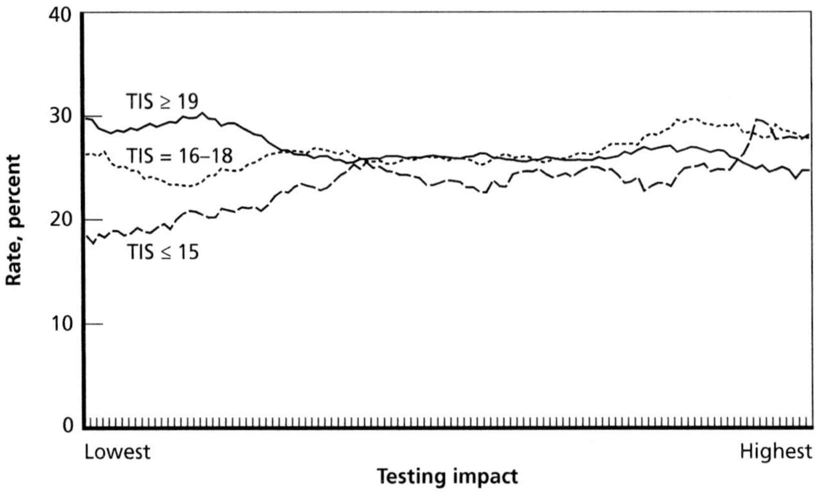

SOURCE: Derived from WAPS history file.
RAND *MG678-4.12*

Figure 4.12 indicates that for E6s with 15 years TIS or less, the Air Force promoted those in AFSCs with the highest testing impact about 10 percentage points higher than their contemporaries in AFSCs with the lowest testing impact.

The 16–18 years TIS group was not as homogenous as the group that had 15 years TIS or less, because it was disproportionately missing good testers from the AFSCs with the highest testing impact (the missing E6s were already promoted to E7 at disproportionately higher rates when they had 15 years TIS or less). As a result, for the 16–18 years TIS group, we only saw about a two-percentage-point difference (= 28 percent – 26 percent) between the selection rates of AFSCs with the highest and lowest testing impacts.

The group with 19 years TIS or more was even less homogenous than the 16–18 years TIS group. For the former, the disproportionately better testers in the AFSCs on the left side of the figure closed the selection rate gap with the poorer testers on the right side.

For the AFSCs on the left, other factors, particularly longevity, had relatively more impact. Hence, on the left, the group with 19 years TIS or more had a higher selection rate than the groups with 16–18 years TIS and with 15 years TIS or less. On the right side of Figure 4.12, the good testers in the group with 15 years TIS or less outpaced the more senior, but poorer, testers in the group with 19 years TIS or more.

Had we not used moving averages, the 15-years-or-less TIS line in Figure 4.12 would have looked like Figure 4.13. Parallel to our analyses of selections to grades E5 and E6, we developed models to help us isolate and quantify the true impact of variations in test scores.

A Multivariate Perspective of Selections to E7

We developed four multivariate linear regression models to predict selections to E7 as functions of TIS and other factors (Appendix H).

Figure 4.13
05E7 Selection Rate Versus Testing Impact, TIS Less Than or Equal to 15

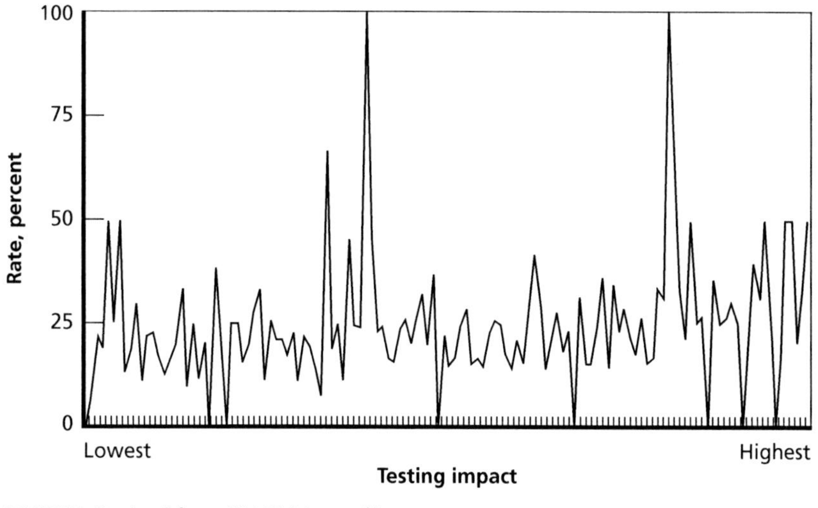

SOURCE: Derived from WAPS history file.
RAND *MG678-4.13*

Figure 4.14 plots modeled selection rates for E6s in AFSCs with the highest and lowest testing impacts. It uses average values for EPR, decorations, and longevity impacts for non-CCS AFSCs.

For AFSCs with the lowest testing impacts, Figure 4.14 shows that selection rates increased with increasing longevity and that the Air Force promoted senior E6s at almost four times the rate of fast burners. However, for those in AFSCs with highest testing impacts, the Air Force selected fast burners at higher rates than it did senior E6s who averaged in excess of 20 additional longevity points.

Selections to E8

We have shown that the Air Force tends to accelerate selections to E5–E7 for junior members in AFSCs or grades that have greater testing impacts. In the following analysis, we show that E8 selection boards do not subsequently penalize junior E7s because they are inexperienced.

Figure 4.14
Modeled Selection Rate Versus Testing Impact, 05E7 Cycle

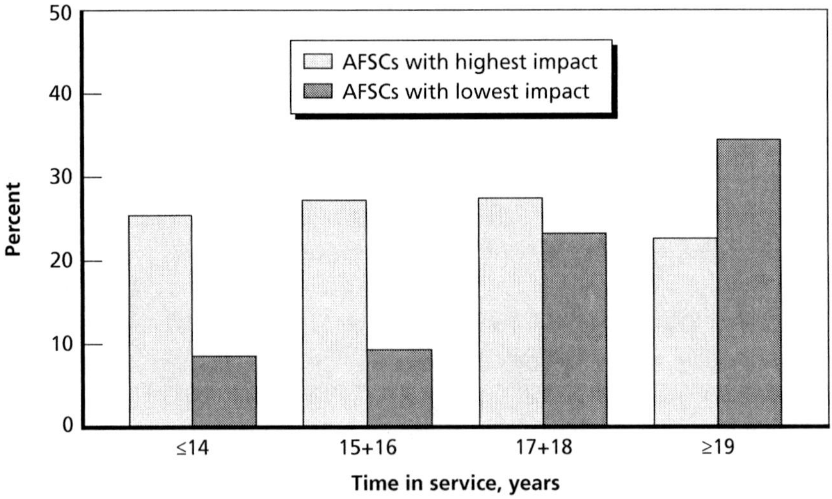

SOURCE: Derived from WAPS history file.
RAND *MG678-4.14*

E8 Selection Factors

We saw in Figure 2.20 that board scores have, by far, the highest impact on E8 cycles. This happens because smaller ranges of testing and longevity points are available; because there is very little variation in EPR scores; and because a range of 180 board score points is available (from a minimum of 270 to a maximum of 450) and boards tend to award the entire range of points available to them.

Since selection rates to E8 tend to be 10 percent or lower, selection to E8 is highly dependent on earning a high board score. Figure 4.15 shows selection rates by board score deciles for the 98E8–05E8 cycles. Over this period, those with board scores in the top ten percent within their AFSCs had a selection rate of 63 percent. In sharp contrast, those with board scores in the bottom 50 percent realized virtually no selections. Therefore, gaining insight into who garners top board scores is one key to understanding E8 selections.

Figure 4.15
E8 Selection Rates Versus Board Score Deciles

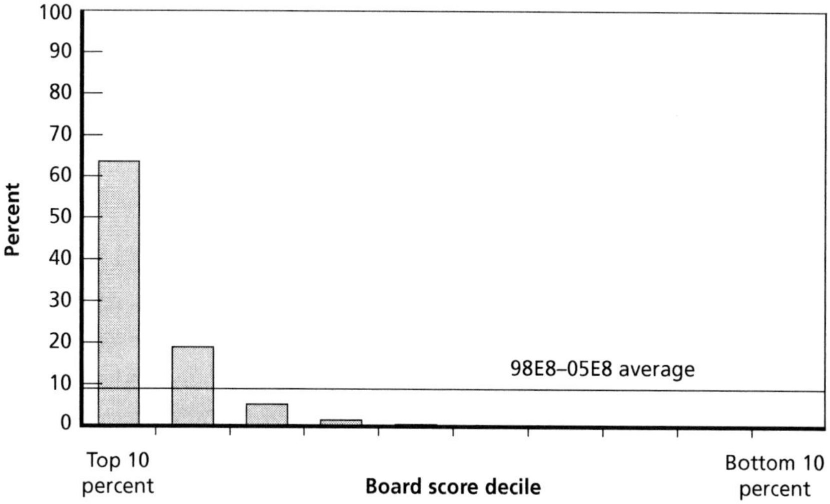

SOURCE: Derived from WAPS history file.
RAND *MG678-4.15*

Figure 4.16 shows the relationship between top board scores and time in service. For the 98E8–05E8 cycles, 25 percent of the E7s competing with 14 years TIS or less had board scores in the top 10 percent. At the other extreme, fewer than 5 percent of those with 26 or more years TIS earned top board scores.[10]

Multiple factors influence board scores. The annual performance reports that E7s–E9s receive may be signed, based on performance, by an individual's senior rater, deputy senior rater, or less than a deputy senior rater. Endorsements by senior raters send strong signals to selection boards that the individuals in question are exceptional performers.

Figure 4.16
Top Board Scores Versus Time in Service

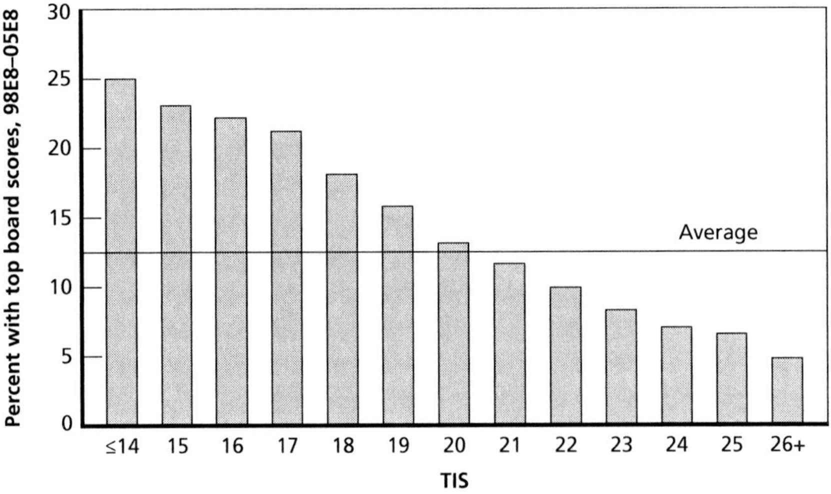

SOURCE: Derived from WAPS history file.
NOTE: The average line exceeds 10 percent because we counted the top 10 percent plus those with tied board scores.
RAND *MG678-4.16*

[10] We do not conclude that there is a cause-and-effect relationship between TIS and board scores. However, it does not appear that boards penalize individuals for having too little TIS.

There is a strong positive relationship between E8 and E9 board scores and the number of consecutive senior rater endorsements.[11] Finally, it is common knowledge within the senior officer and senior NCO ranks that board scores are also related to consistent, compelling stratification in the written portions of EPRs.[12]

Selections to E9

Figure 4.17 shows selection rates by board-score deciles for the 98E9–05E9 cycles. Over this period, those with board scores in the top 20

Figure 4.17
E9 Selection Rates Versus Board Score Deciles

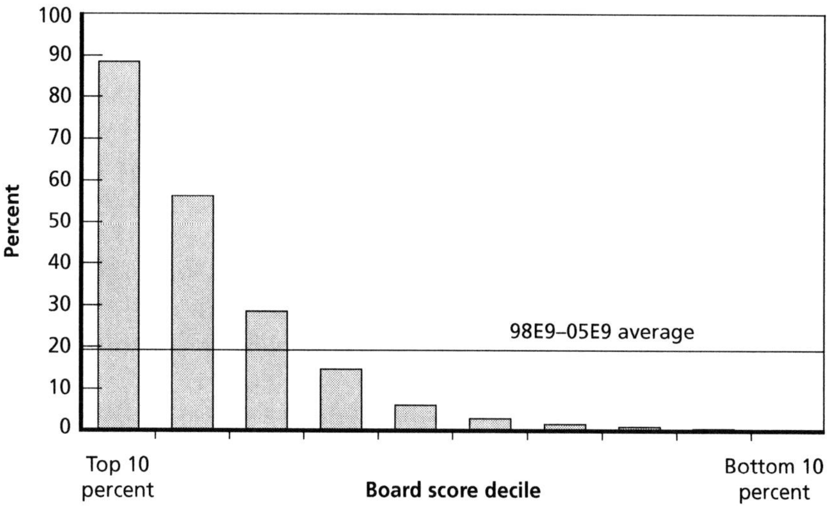

SOURCE: WAPS history file.
RAND *MG678-4.17*

---

[11] See Moore, 1998, p. 2.

[12] *Stratification* compares an individual to his/her immediate peers. For example, "#1 of 151 E7s in the wing" would be strong stratification. "#2 of 7 senior NCOs in the branch" is less compelling

percent in their AFSCs had a selection rate of 72 percent. In sharp contrast, those with board scores in the bottom 50 percent realized a selection rate of 1 percent. Therefore, gaining insight into which E8s receive top board scores is a key to understanding E9 selections.

Figure 4.18 shows the correlation between top board scores and time in service. For the 98E9–05E9 cycles, 48 percent of those competing with 17 years TIS or less received board scores that were in the top 20 percent in their AFSCs. At the other extreme, only 13 percent of those with 27 or more years TIS earned board scores in the top 20 percent.

Chapter Summary

In this chapter, we have demonstrated graphically and with models that junior E4s–E6s tend to get selected at higher rates when they are

Figure 4.18
E9 Board Scores in Top 20 Percent Versus TIS

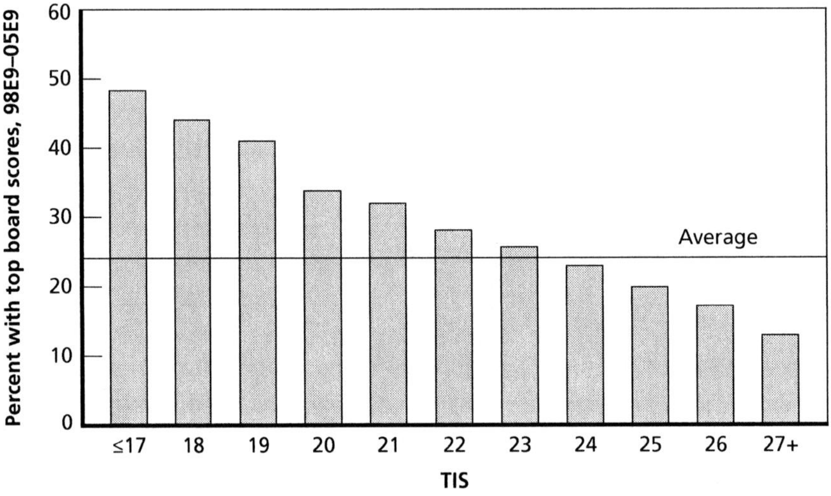

SOURCE: WAPS history file.
RAND *MG678-4.18*

in AFSCs that have higher testing impacts. E8 and E9 selection boards do not subsequently penalize junior E7s and E8s who were able to progress rapidly through the enlisted ranks.

CHAPTER FIVE

Effects

This chapter builds on Chapter Four, which established that differences in testing impacts explain differences in selection rates within year groups, especially to the grades of E5–E7. In addition to our concerns about achieving AFPD 36-25 objectives, we also believe that the differences in testing impact across AFSCs produce forcewide effects that the Air Force needs to recognize and potentially manage. We show that the policy of not standardizing test scores influences senior NCO manning. In turn, this influences the Air Force's candidate pool for nominative and commander-involvement chief master sergeant positions. Finally, it also means that equally experienced individuals in different AFSCs do not have the same opportunity to achieve senior NCO status, which violates the intent of those who designed the enlisted promotion system.

Inconsistent and Random Selection Standards

We demonstrated in Chapter Four that the Air Force promotes fast burners earlier in some AFSCs because of differences in testing impacts. Unintentionally applying different selection standards that randomly vary over time means that the Air Force cannot know if it is selecting the most qualified people, especially to the grades of E5–E7.

Senior NCO (E7–E9) Manning

Suppose two AFSCs have the same number of E6 authorizations and that force managers consistently achieve 100 percent E6 manning for both. Under ESO, both would generate about the same number of E7 selections annually. We know from Figure 2.15 that the two AFSCs would very likely have WAPS tests with different impacts on selection outcomes; in this section, we assume that they are different. We know from Chapter Four that, all other things being equal, the AFSC with the greater testing impact will produce younger E7s. In turn, we know that the AFSC with the earlier phase point will also produce more E7s–E9s because these senior NCOs are able to serve for longer periods before reaching mandatory retirement milestones. The ability to serve longer produces a disproportionately larger base of eligibles, who then reap a larger share of available selections under ESO.

As we debated this notion internally, we questioned whether an AFSC could consistently sustain lower phase points to any grade. Some members of the team postulated that, because WAPS awards individuals additional points for seniority, it might be self-correcting. The thought was that, even if testing in an AFSC had a large impact that kept phase points low initially, the numbers of senior E6s in the AFSC would grow and the Air Force would eventually promote them. In turn, their seniority would increase phase points. Because of the dynamic nature of this hypothesis, we turned to simulation.

Our simulations did reveal that phase points do initially oscillate. However, as we let our simulated systems reach a steady state, we observed the results that we predicted. Figure 5.1 plots the ratio of an AFSC's E7s through E9s to its E6s against the standard deviation of test scores (PFE + SKT) for selection to E7, all other things being equal. The figure implies that a fixed number of E6s produces increasing numbers of senior NCOs as the variance of its test scores increases.

While our simulations produced about the same range of ratios that we saw in Figure 1.2, the magnitudes of our ratios are greater.

Figure 5.1
Simulation Results

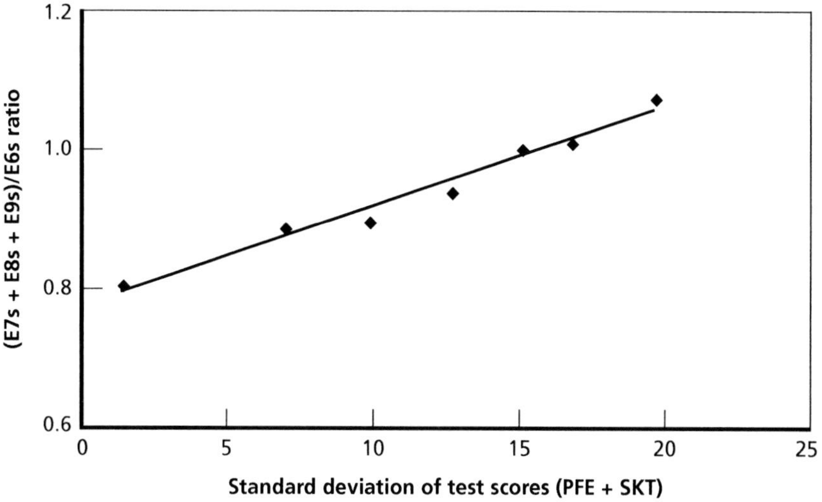

RAND *MG678-5.1*

We attribute these differences to the fact that our simulation did not migrate individuals out of their AFSCs into tax AFSCs.

Depending upon an AFSC's requirements in the top three grades, producing larger or smaller top three inventories is not necessarily bad. However, because the Air Force does not standardize scores, testing variations currently produce random manning effects.

Unequal Opportunities to Make E8 and E9

From the perspective of individuals, Figure 5.1 also means that members of AFSCs with greater testing impacts have better chances of making E8 and E9. This is not consistent with the Air Force's policy of equal promotion opportunity, which the Air Force first embraced through TOPCAP.

Disproportionate Selectivity for E9 Nominative and Commander-Involvement Positions

Because the policy of not standardizing test scores produces differences in selection timing for AFSCs, we believe that chief master sergeants from some AFSCs have a reduced opportunity to hold strategic E9 jobs.

The Air Force uses either nomination or commander-involvement processes to identify chief master sergeants to fill about 400 positions with strategic leadership or management responsibilities.[1] Typically, the Air Force fills these positions with chiefs who have performed well in previous jobs as chiefs. However, filling multiple jobs as a chief before reaching mandatory retirement at 30 years of service dictates that some individuals make E9 at 20 years of service or sooner. In turn, this requires that the Air Force promote some individuals to E8, E7, E6, and E5 well ahead of their peers. For example, Figure 5.2 shows

Figure 5.2
Phase Points to E7 for Strategic Chiefs

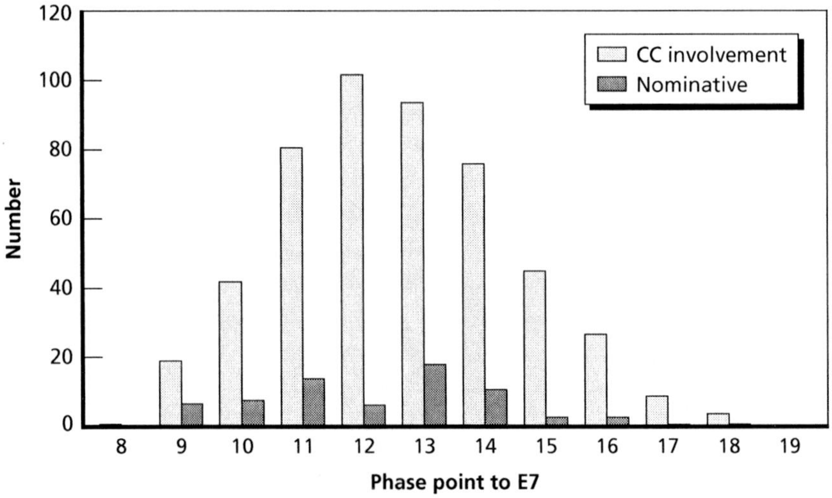

SOURCE: Derived from AFPC WAPS history file.
RAND *MG678-5.2*

---

1 An example of a strategic chief job is MAJCOM Command Chief.

the distribution of phase points to E7 for chiefs who recently served in strategic E9 jobs. The Air Force promoted the majority of these chiefs (84 percent) to E7 with 14 or fewer years of service.

Figure 5.3 shows that these chiefs became E7s primarily in the first half of the 1990s. In contrast, during this same period, the Air Force average phase point to E7 was about 16 years of service (Figure 5.4). Hence, the Air Force promoted almost all chiefs filling nominative or commander-involvement positions to E7 at least two years ahead of the average, and it promoted over half of them four or more years before their contemporaries.

Tying these facts to the discussion of differential testing impacts, we can conclude that, by promoting fast-burner E6s later when they are in AFSCs that have lower testing impacts, the Air Force restricts its pool of candidates to fill future strategic chief positions. We believe that differences in testing impacts explain Figure 1.2, which shows that at the 2-digit level, AFSCs have wide variations in percentages of

Figure 5.3
Date of Rank to E7 for Recent Strategic Chiefs

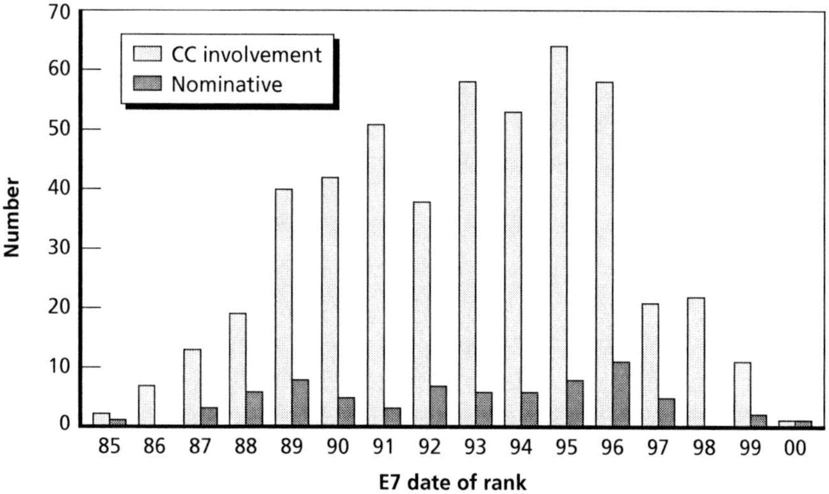

SOURCE: Derived from AFPC personnel data system.
RAND *MG678-5.3*

Figure 5.4
Average Phase Points to E7 by Fiscal Year

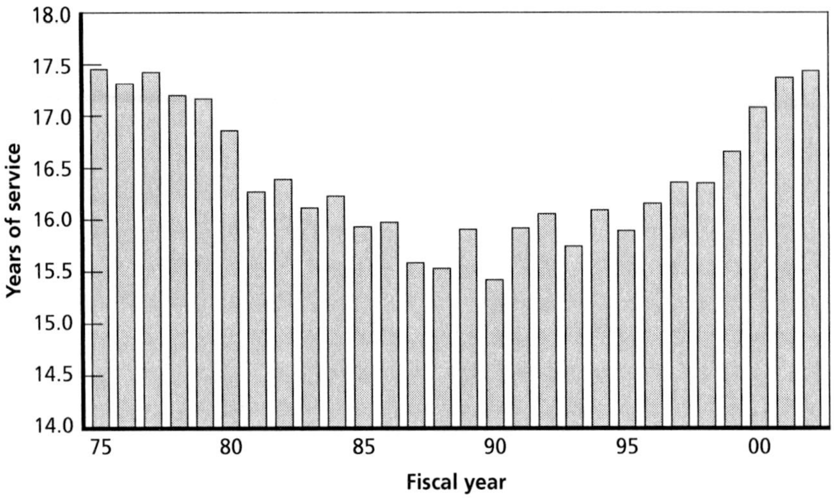

SOURCE: Derived from AFPC military personnel data system.
RAND *MG678-5.4*

E9s with 20 or fewer years of service. This situation may not be in the Air Force's best interest.[2]

Standardization Strategies

In Figures 3.2 and 3.3, we transformed raw scores into standardized scores that had standard deviations of 11. In practice, the Air Force could regulate the impact of testing by judiciously selecting the variances of standardized scores. To establish standard deviations, the Air Force would need to examine its priorities. The first question is whether achieving promotion equity (as opposed to selection equity) is a pri-

---

[2] Unpublished RAND survey results indicate that 63 percent of commander-involvement and nominative chief positions require experience in a specific AFSC (or set of AFSCs). For the remaining 37 percent, one must wonder if Air Force leadership is not better served by multiple functional perspectives. There should also be motivational benefits when young airmen in an AFSC observe one of their own people filling a strategic chief position.

mary objective. If it is, the variation in standardized scores should be the same for every AFSC. Making all standard deviations equal would be computationally trivial. In addition, it would greatly reduce, but not eliminate, the differences in testing impact across AFSCs. As we saw in previous chapters, testing impact is also a function of the variations in the other WAPS factors. If the Air Force wished to adopt a standardization approach to equate testing impacts, it would be possible to tune the variance in each AFSC's standardized scores so that testing had the same impact in every AFSC.

It might be the case that satisfying authorizations is more important to the Air Force than achieving promotion equity. By deliberately calibrating the variances of standardized scores (and the impact of testing), the Air Force could influence the number of senior NCOs in each AFSC.

The Air Force could also use test score standardization to regulate the numbers it promotes to chief master sergeant before 20 years of service. In turn, this would control the eventual number of candidates for nominative and commander-involvement positions. Using current retention and selection rates, Figure 5.5 shows our simulation results that estimate the relationship between the standard deviation of test scores and the percentage of E9s with 20 or fewer years of service.[3] Because loss rates, selection rates, and phase points are dynamic, analysts would periodically need to recalculate the desired standard deviations for AFSCs and grades.

The simulated results illustrated in Figure 5.5 reflect higher percentages of deep-selected E9s than we observed in the data (see Figure 1.2).[4] The Air Force's enlisted assignment system tends to reduce the number of deep-selects in AFSCs because of the requirement to assign individuals to tax AFSCs. To the extent that any deep-selects retrain into tax AFSCs, the losing AFSCs will reflect lower percentages of deep-selects. The Air Force Personnel Center, when it identifies indi-

---

[3] We ran entity level simulations. The attributes of our system and the individuals in our simulations mirrored reality in 2005. Our simulations accounted for key personnel management policies, e.g., HYT.

[4] *Deep-selected* individuals have phase points (to chief in this case) well below average.

Figure 5.5

Relationship Between Standard Deviation of Test Scores and Deep-Selected E9s

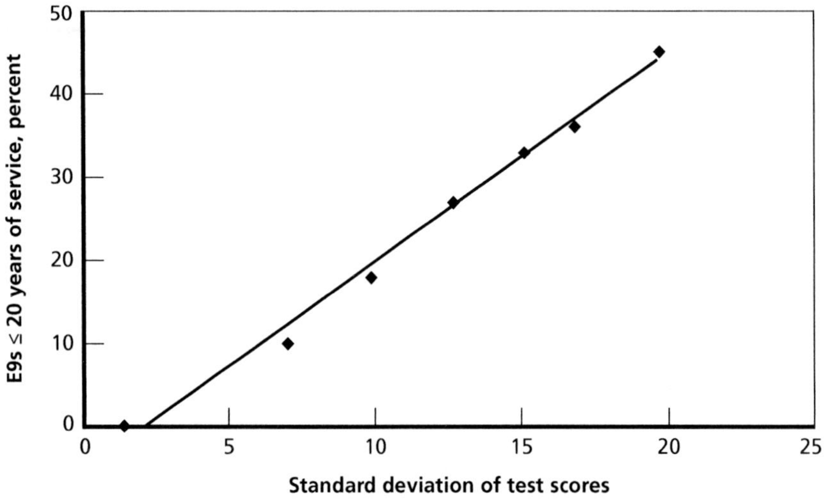

RAND *MG678-5.5*

viduals for tax assignments, could also be disproportionately tapping overmanned AFSCs. Again, our simulation results indicate what would occur in an environment without tax assignments. The main message from Figure 5.5 is that there is a positive relationship between the standard deviations of test scores and the percentage of an AFSC's E9s who are deep-selects. Therefore, there is a positive relationship between the standard deviations of test scores and the opportunity for chiefs in an AFSC to serve in strategic positions.

While African-Americans tend to get selected at about average rates, their phase points to E5–E9 generally average about six to 12 months greater than the Air Force mean. This is because their PFE and SKT scores tend to be below average, and this can be tied to lower AFQT scores. African-Americans are also overrepresented in some AFSCs.[5] To the extent that African-Americans or any other group with

5 Fuller, 2001.

lower test scores are in AFSCs that have higher testing impacts, phase point differences are magnified. Conversely, where testing has lower impacts, phase point differences are reduced. Hence, when selecting a standardization strategy, the Air Force might wish to be mindful of the implications for minorities.

Transition Issues

If the Air Force were to standardize test scores, it would need to consider implementation timing. One option would be to standardize testing scores in a single selection cycle. Figure 5.6 shows what might have happened if the Air Force had adopted this implementation approach for the 05E5–05E9 cycles, and it assumes that achieving promotion equity is the objective.

The analysis behind Figure 5.6 assumes that the standard deviation of PFE and SKT scores for each AFSC/grade should equal the standard deviation of scores for all testers combined. Thus, this assumption draws AFSCs with very high and very low weights toward the center.

Figure 5.6
Potential Single-Cycle Impact of Standardization on Individuals

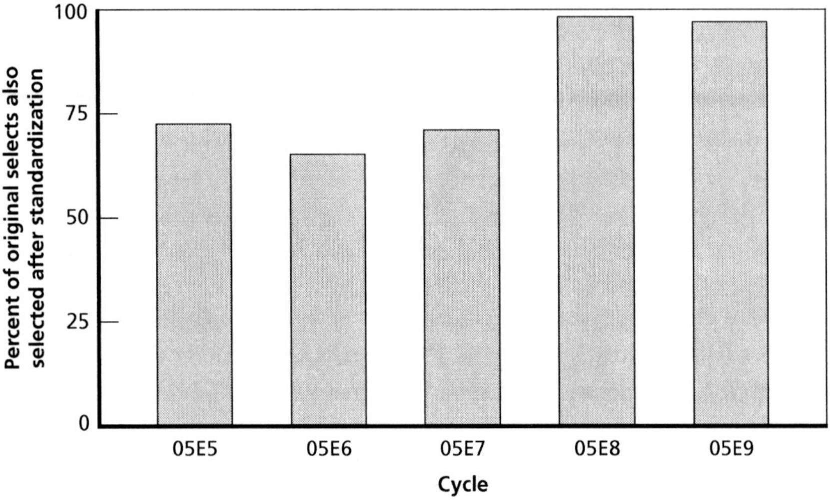

RAND *MG678-5.6*

Each bar in Figure 5.6 shows the percentage of actual selects who would have still been selected in that cycle using test scores standardized in this way. Overall, about 75 percent of the E5–E7 selects would have remained unchanged.[6] At the AFSC level of detail, on the 05E5 cycle, the AFSC that would have experienced the greatest standardization impact would have been AFSC 3P0X1, Security Forces. In this example, 43 percent of those actually selected would have been selected the first year of standardization. Security Forces would have also been the most heavily affected AFSC on the 05E6 cycle, with 34 percent of the actual selects also being selected using standardized scores. For both the E5 and E6 cycles, testing had a very low impact in Security Forces selections, but that impact would increase under standardization. Under standardization, we would expect the AFSC 3P0X1 E5

[6] We standardized variances only. After substituting the standardized test scores for the raw scores, we reordered the eligibles and drew the cut lines based on the original number of selects in each AFSC. From the perspective of the original non-selects, 11 percent would have been selected on the 05E5 cycle, 7 percent on the 05E6 cycle, 8 percent on the 05E7 cycle, 0.2 percent on the 05E8 cycle, and 0.8 percent on the 05E9 cycle.

and E6 selects to be younger. In turn, Security Forces would eventually produce more in the top three grades and more deep-selected chiefs.

Another key insight from Figure 5.6 is that standardizing test scores would have little impact on the E8 and E9 selection process, which would continue to be dominated by board scores.

The results in Figure 5.6 represent but one of many standardization implementation strategies the Air Force might adopt. If the Air Force wished to increase or decrease the number of young chiefs by increasing or decreasing the average impact of testing for the E5–E7 cycles, we could anticipate seeing greater differences in the single-cycle selects.

Standardization Costs

If the Air Force decided to standardize test scores, there would be three basic types of costs: implementation costs, marketing costs, and maintenance costs. The implementation costs would have software and analysis components. The software changes should be minimal and as simple as applying Equation 3.1 to convert raw scores to standardized scores. The system that the Air Force currently uses to report raw scores to individuals might just as well be reporting standardized scores. The primary implementation costs would be associated with determining the variances for standardized scores. One approach would be to capture the policies of board members as they evaluated E4, E5, and E6 records (there should already be enough data to capture the policies of boards that have already evaluated E7 and E8 records). A better approach would probably be to first establish policy objectives. For example, the Air Force might establish target distributions of phase points to each grade. The next step would be to develop a high-fidelity enlisted selection model that could predict the implications of setting standardized variances at various levels. We estimate that the up-front analytical costs for this more deliberate approach could be three to four person-years.

It would be appropriate for the Air Force to market any shift to standardized scores. To limit the loss of confidence in the current system, it would have to give some thought to the message. The infrastructure already exists to allow the leadership to communicate with the enlisted force, so the primary cost would be designing the message.

Finally, to keep variances in standardized test scores effectively related to policy objectives, the Air Force would need to commit to continuing enlisted analysis support to develop annual targets for the variation in standardized test scores. Depending on its standardization strategy, this effort might consume up to one person-year annually.

CHAPTER SIX

Conclusions and Recommendation

Conclusions

Because the Air Force does not standardize PFE and SKT scores, testing does not play the same role in determining selection outcomes in every grade or AFSC or within AFSCs over time. Because these differences are not by design, the Air Force is experiencing random WAPS factor impacts in promotion selections.

Differential impacts of testing raise promotion equity concerns. When AFSCs have large ranges in test scores, their young, good testers will advance more rapidly. When large testing point spreads are consistent over time and across grades, AFSCs will actually produce more senior NCOs from a given number of accessions than will AFSCs with low variations in test scores. For the same reason that some AFSCs produce more chiefs, they also produce higher percentages of young chiefs. In turn, this affects the pool from which the Air Force selects chiefs to fill commander-involvement and nominative jobs.

Recommendation

Before the Air Force makes a decision about standardizing test scores, it must first decide whether or not it wishes to fully achieve the primary objective established in AFPD 36-25. The current approach to testing is producing random effects that guarantee that the AF cannot be sure that it is promoting individuals with the highest potential to the grades of E5–E7. If the Air Force wishes to meet this objective, it must standardize test scores.

Assuming that the Air Force standardizes scores, the next step would be to define what AFPD 36-25 means by "highest potential to fill positions of increased grade and responsibility." To that end, the first question to ask is whether or not the Air Force should still pursue the TOPCAP goal of equal promotion opportunity (to include equal phase points) for each AFSC. If so, the Air Force needs to better equalize the impacts of testing across AFSCs. Establishing the magnitudes of standardized test score variations would implicitly define "highest potential" and the desired relationships between testing and longevity. For example, if the Air Force needs to proportionally produce more young chiefs from all AFSCs so that it will subsequently have greater selectivity when it fills commander-involvement and nominative positions, it should take measures to increase the variances in test scores to a uniform target.

Since the Air Force has never achieved the promotion equity envisioned by TOPCAP and because the NCO force is nonetheless very robust in the aggregate, one could argue that promotion equity should not be a concern. If achieving equal promotion opportunity is not a constraint and if there is a desire to maintain the policy of ESO, the Air Force might wish to define "highest potential" differently for each AFSC. Is experience more valued than testing ability in some AFSCs? For AFSCs with higher top-three/E6 requirement ratios, should testing have a greater effect on selection outcomes so that those AFSCs can produce larger top three inventories? Controlling the E7–E9 inventory within an AFSC by controlling the variation of test scores should only be a long-term management option in the presence of stable requirements. However, top-three grade requirements tend not to be stable under the current decentralized requirements determination process. In addition, as we discuss at length in *Air Force Enlisted Force Management: System Interactions and Synchronization Strategies* (Schiefer et al., 2007), the top three requirements for many AFSCs are not executable under the current policy of equal selection opportunity. We make the case that matching inventory to requirements and achieving equal selection opportunity need not be mutually exclusive if the Air Force would centrally constrain the requirements determination process to ensure that it does not establish unachievable goals.

It is not clear which office should take the lead on the standardization question. A number of parties have legitimate and sometimes competing interests, but no single office is responsible for enlisted management. Ultimately, the decision should probably be a function of the policy objectives that the Air Force seeks to achieve by standardizing test scores. It is clear that the Air Force would need to make an unwavering analysis investment. Because the distributions of points earned through other WAPS factors are not static, analysts would need to periodically derive target variations in standardized test scores. Even if the Air Force decides to seek a surrogate for standardization by modifying the techniques that the Occupational Measurement Squadron uses to develop tests, that group would still require initial and reoccurring guidance on target variances.

We close by encouraging the enlisted managers in the Air Force to make deliberate policy choices on the testing issues we have surfaced. Today, the well-intentioned senior NCOs who develop WAPS tests are implicitly making decisions that rightfully belong to the Air Staff.

APPENDIX A

The Impact of WAPS Factors by Grade and AFSC

The tables in this appendix show the average impacts of WAPS factors in stable, non-tax AFSCs for the selection cycles conducted from 1998 through 2005, for AFSCs with at least 25 eligibles per cycle. These impacts were determined using Approach Four as described in Chapter Two.[1] Tables A.1 through A.3 provide these impacts for selections to grades E5 through E7, respectively. The AFSCs in the tables are sorted by increased testing impact. For reference, Appendix B provides the titles of the AFSCs appearing in these tables.

Table A.1
Average Impacts of WAPS Factors for 87 Stable AFSCs, 98–05 E5 Cycles

	Average Impacts			
AFSC	EPRs	Decorations	Testing	Longevity
2T3X1	0.51	0.03	0.54	0.33
2T0X1	0.41	0.03	0.55	0.30
3E7X1	0.49	0.03	0.55	0.29
1N6X1	0.43	0.03	0.55	0.30
2T3X5	0.42	0.03	0.55	0.36
3A0X1	0.43	0.04	0.55	0.35
1C4X1	0.42	0.05	0.56	0.26

[1] To calculate impact for a factor in an AFSC, Approach Four calculates the average change in standard deviation from the mean when that factor is removed from the WAPS formula. Hence, an impact of 0.60 for testing in an AFSC means that, on average, each individual would move 0.6 standard deviations within the AFSC when we ignore test scores.

Table A.1—continued

AFSC	Average Impacts			
	EPRs	Decorations	Testing	Longevity
3S0X1	0.42	0.04	0.56	0.35
2A7X3	0.50	0.02	0.57	0.34
4Y0X1	0.51	0.02	0.57	0.33
4B0X1	0.44	0.03	0.57	0.34
2W1X1	0.42	0.03	0.57	0.31
3E2X1	0.40	0.04	0.57	0.30
2A6X2	0.49	0.02	0.57	0.31
4A1X1	0.46	0.03	0.58	0.32
3M0X1	0.45	0.03	0.58	0.33
2A6X4	0.50	0.02	0.58	0.30
5R0X1	0.52	0.04	0.58	0.34
1T1X1	0.39	0.03	0.58	0.28
2A7X1	0.51	0.02	0.58	0.36
1A0X1	0.38	0.15	0.59	0.30
3C2X1	0.41	0.03	0.59	0.35
3E4X1	0.42	0.03	0.59	0.34
2A6X1B	0.46	0.02	0.59	0.32
4P0X1	0.51	0.02	0.59	0.35
4A0X1	0.45	0.03	0.59	0.35
3E3X1	0.40	0.04	0.59	0.36
2S0X1	0.42	0.03	0.59	0.35
2A6X5	0.45	0.03	0.59	0.34
2A7X4	0.46	0.02	0.60	0.32
2T2X1	0.43	0.03	0.60	0.32
2A3X1B	0.38	0.02	0.60	0.29
2A6X1A	0.44	0.02	0.60	0.33
2A7X2	0.42	0.02	0.60	0.33
2A3X3B	0.46	0.02	0.60	0.34
4C0X1	0.39	0.02	0.60	0.34
2A3X3J	0.44	0.03	0.60	0.32
4N0X1	0.46	0.02	0.60	0.35
3P0X1	0.41	0.03	0.60	0.29
4N1X1	0.48	0.02	0.60	0.29
2T1X1	0.43	0.03	0.60	0.31
4E0X1	0.44	0.03	0.61	0.31
3V0X2	0.40	0.04	0.61	0.42
1C0X2	0.44	0.03	0.61	0.32

Table A.1—continued

	Average Impacts			
AFSC	**EPRs**	**Decorations**	**Testing**	**Longevity**
2W2X1	0.43	0.03	0.61	0.31
2A6X3	0.47	0.02	0.61	0.28
2M0X3	0.39	0.02	0.61	0.34
2A3X1C	0.44	0.03	0.61	0.31
2F0X1	0.39	0.03	0.61	0.32
2R1X1	0.37	0.03	0.61	0.32
2A3X3A	0.43	0.03	0.61	0.31
3E1X1	0.41	0.03	0.61	0.34
2W0X1	0.44	0.03	0.61	0.34
3E0X2	0.41	0.04	0.62	0.30
6C0X1	0.39	0.03	0.62	0.35
1C1X1	0.37	0.03	0.62	0.30
3C0X1	0.39	0.03	0.62	0.34
2A6X6	0.42	0.02	0.62	0.33
3C1X1	0.40	0.03	0.62	0.31
3E0X1	0.40	0.04	0.62	0.34
1C3X1	0.40	0.03	0.62	0.28
6F0X1	0.40	0.03	0.62	0.33
1N0X1	0.39	0.04	0.63	0.31
2A3X1A	0.43	0.03	0.63	0.30
2A5X1L	0.40	0.02	0.63	0.32
1N1X1	0.40	0.06	0.63	0.29
2A5X1J	0.40	0.03	0.64	0.33
4T0X1	0.46	0.02	0.64	0.38
2R0X1	0.37	0.03	0.64	0.33
2E1X2	0.36	0.03	0.64	0.37
2M0X1	0.35	0.02	0.64	0.33
3C0X2	0.39	0.03	0.65	0.35
2A0X1A	0.40	0.02	0.65	0.30
2E6X3	0.43	0.03	0.65	0.36
2E6X2	0.43	0.04	0.65	0.35
2A5X1K	0.40	0.02	0.66	0.32
3E8X1	0.34	0.04	0.66	0.34
2E1X4	0.44	0.04	0.66	0.37
1N4X1	0.31	0.03	0.67	0.31
3V0X1	0.39	0.04	0.67	0.36
2E2X1	0.38	0.03	0.68	0.33

Table A.1—continued

	Average Impacts			
AFSC	EPRs	Decorations	Testing	Longevity
4R0X1	0.48	0.02	0.68	0.35
1C6X1	0.37	0.03	0.68	0.31
1N2X1	0.38	0.02	0.70	0.36
2E1X1	0.38	0.03	0.71	0.34
2E0X1	0.39	0.03	0.71	0.38
2E1X3	0.39	0.04	0.75	0.39

Table A.2
Average Impacts of WAPS Factors for 103 Stable AFSCs, 98–05 E6 Cycles

	Average Impacts			
AFSC	EPRs	Decorations	Testing	Longevity
3E4X3	0.11	0.08	0.49	0.51
2E1X4	0.15	0.09	0.53	0.55
2T3X7	0.24	0.10	0.53	0.52
7S0X1	0.12	0.11	0.55	0.52
3N0X1	0.20	0.10	0.55	0.48
2P0X1	0.22	0.08	0.55	0.49
3C3X1	0.18	0.09	0.55	0.51
4A1X1	0.21	0.08	0.56	0.49
3E9X1	0.21	0.11	0.56	0.57
1A5X1	0.13	0.20	0.56	0.40
3C2X1	0.18	0.09	0.56	0.58
5R0X1	0.19	0.10	0.56	0.52
2E0X1	0.20	0.09	0.57	0.58
4P0X1	0.23	0.07	0.57	0.55
2A0X1A	0.22	0.07	0.57	0.49
2A0X1B	0.16	0.08	0.57	0.48
2R1X1	0.17	0.08	0.57	0.50
4B0X1	0.23	0.08	0.57	0.46
3E5X1	0.12	0.09	0.57	0.48
4M0X1	0.12	0.08	0.57	0.43
2A3X1	0.19	0.09	0.58	0.59
3E7X1	0.25	0.08	0.58	0.51
2T3X0	0.23	0.10	0.58	0.56
2G0X1	0.15	0.11	0.58	0.51
2A6X1A	0.18	0.08	0.59	0.51

Table A.2—continued

	Average Impacts			
AFSC	EPRs	Decorations	Testing	Longevity
2E1X3	0.18	0.11	0.59	0.54
4Y0X1	0.23	0.07	0.59	0.52
4D0X1	0.31	0.07	0.59	0.60
6C0X1	0.17	0.09	0.59	0.55
4N1X1	0.22	0.07	0.59	0.51
2W0X1	0.20	0.08	0.59	0.52
1C4X1	0.21	0.14	0.59	0.51
2A5X2	0.19	0.09	0.59	0.51
4N0X1	0.24	0.09	0.59	0.53
3E4X1	0.16	0.09	0.59	0.50
3C0X2	0.21	0.10	0.59	0.63
3S2X1	0.18	0.09	0.59	0.54
3S0X1	0.15	0.09	0.59	0.50
3A0X1	0.15	0.10	0.60	0.52
2M0X1	0.16	0.07	0.60	0.51
2W1X1	0.18	0.09	0.60	0.51
2E1X1	0.19	0.10	0.60	0.57
3E3X1	0.15	0.10	0.60	0.50
6F0X1	0.16	0.09	0.60	0.47
2A7X3	0.21	0.08	0.60	0.55
2A3X2	0.21	0.09	0.60	0.58
2T0X1	0.16	0.09	0.60	0.46
2E2X1	0.17	0.08	0.60	0.52
2S0X1	0.17	0.08	0.60	0.47
4A2X1	0.14	0.08	0.60	0.43
4E0X1	0.21	0.08	0.60	0.46
2E6X2	0.16	0.08	0.60	0.51
3V0X2	0.19	0.10	0.60	0.55
4A0X1	0.21	0.08	0.61	0.51
1W0X1A	0.25	0.10	0.61	0.56
3C0X1	0.19	0.10	0.61	0.54
2A6X6	0.18	0.09	0.61	0.56
2A5X3A	0.15	0.09	0.61	0.54
1T1X1	0.19	0.07	0.61	0.47
2E6X3	0.15	0.09	0.61	0.50
1N1X1	0.19	0.12	0.61	0.46
3E0X2	0.16	0.10	0.61	0.50
2A6X4	0.21	0.08	0.61	0.51

Table A.2—continued

AFSC	Average Impacts			
	EPRs	Decorations	Testing	Longevity
1C5X1	0.17	0.09	0.61	0.46
2A6X3	0.20	0.08	0.62	0.54
1A3X1	0.15	0.21	0.62	0.46
2A6X5	0.19	0.09	0.62	0.50
1A4X1	0.15	0.19	0.62	0.41
2F0X1	0.19	0.08	0.62	0.48
3P0X1	0.17	0.09	0.62	0.48
2T2X1	0.20	0.09	0.62	0.50
2T1X1	0.23	0.09	0.62	0.50
2A5X1	0.18	0.09	0.62	0.56
3E8X1	0.13	0.10	0.62	0.47
4Y0X2	0.20	0.07	0.62	0.48
3E2X1	0.15	0.09	0.62	0.46
2A7X1	0.24	0.08	0.62	0.52
3E1X1	0.14	0.09	0.62	0.52
1C1X1	0.18	0.08	0.63	0.48
2A6X1B	0.22	0.08	0.63	0.55
3M0X1	0.24	0.10	0.63	0.53
3E0X1	0.11	0.09	0.63	0.48
5J0X1	0.12	0.09	0.63	0.47
2M0X2	0.14	0.08	0.63	0.49
4C0X1	0.20	0.08	0.63	0.54
2A7X4	0.22	0.07	0.64	0.50
1N0X1	0.16	0.12	0.64	0.53
1N2X1	0.16	0.10	0.64	0.46
2R0X1	0.17	0.09	0.64	0.58
4T0X1	0.25	0.08	0.64	0.52
2W2X1	0.18	0.08	0.65	0.49
3V0X1	0.18	0.09	0.65	0.61
1C0X2	0.20	0.10	0.65	0.54
2A6X2	0.21	0.08	0.65	0.54
1N4X1	0.15	0.09	0.66	0.49
1C3X1	0.17	0.09	0.66	0.47
1A0X1	0.12	0.21	0.66	0.41
2M0X3	0.13	0.07	0.66	0.48
1T0X1	0.08	0.09	0.66	0.40

Table A.2—continued

	Average Impacts			
AFSC	EPRs	Decorations	Testing	Longevity
4R0X1	0.24	0.06	0.66	0.50
1N5X1	0.19	0.11	0.67	0.58
1A2X1	0.10	0.20	0.67	0.41
1C6X1	0.15	0.09	0.68	0.47

Table A.3
Average Impacts of WAPS Factors for 84 AFSCs, 98–05 E7 Cycles

	Average Impacts			
AFSC	EPRs	Decorations	Testing	Longevity
1S0X1	0.09	0.16	0.70	0.33
7S0X1	0.14	0.15	0.73	0.48
4P0X1	0.16	0.10	0.75	0.60
3N0X1	0.09	0.15	0.75	0.47
1N3X0	0.14	0.30	0.76	0.48
4T0X1	0.23	0.13	0.79	0.54
3E9X1	0.09	0.10	0.80	0.37
2P0X1	0.15	0.13	0.80	0.50
1C6X1	0.11	0.11	0.80	0.33
2A6X1A	0.14	0.11	0.81	0.41
2A6X1B	0.16	0.11	0.81	0.48
3E3X1	0.11	0.14	0.82	0.48
3C3X1	0.21	0.14	0.82	0.53
1N4X1	0.16	0.15	0.82	0.42
3C2X1	0.22	0.17	0.82	0.45
3S2X1	0.20	0.13	0.82	0.49
6F0X1	0.15	0.14	0.82	0.45
4Y0X1	0.23	0.11	0.82	0.47
2T1X1	0.12	0.14	0.83	0.39
4C0X1	0.14	0.13	0.83	0.44
2E0X1	0.21	0.15	0.83	0.50
1C3X1	0.12	0.14	0.83	0.41
2A5X2	0.19	0.11	0.83	0.45
3E7X1	0.17	0.11	0.84	0.38
3E4X1	0.13	0.12	0.84	0.46

Table A.3—continued

	Average Impacts			
AFSC	EPRs	Decorations	Testing	Longevity
1N0X1	0.14	0.17	0.85	0.54
2S0X1	0.12	0.12	0.85	0.40
2G0X1	0.05	0.19	0.85	0.39
4B0X1	0.18	0.15	0.85	0.40
6C0X1	0.14	0.16	0.86	0.47
2T2X1	0.15	0.13	0.86	0.40
3C0X1	0.13	0.13	0.86	0.43
3C0X2	0.13	0.15	0.86	0.48
2A6X3	0.13	0.11	0.86	0.40
2F0X1	0.11	0.12	0.86	0.40
4D0X1	0.26	0.13	0.86	0.41
1C5X1	0.10	0.11	0.87	0.31
3M0X1	0.16	0.13	0.87	0.36
3S0X1	0.12	0.15	0.87	0.47
3A0X1	0.10	0.15	0.87	0.43
2A7X3	0.20	0.10	0.88	0.42
2E1X4	0.17	0.13	0.88	0.43
2W1X1	0.12	0.14	0.88	0.42
2E1X3	0.15	0.13	0.88	0.40
3S1X1	0.15	0.12	0.89	0.35
2E6X3	0.14	0.13	0.89	0.43
2T0X1	0.15	0.13	0.89	0.36
4N0X1	0.22	0.15	0.89	0.45
1T1X1	0.10	0.09	0.89	0.34
2E1X1	0.14	0.16	0.89	0.43
2A6X6	0.17	0.12	0.89	0.43
2A6X2	0.12	0.12	0.90	0.47
2A5X1	0.12	0.10	0.90	0.39
1C0X2	0.10	0.14	0.90	0.52
4R0X1	0.15	0.10	0.90	0.44
2T3X7	0.16	0.15	0.91	0.52
4Y0X2	0.16	0.09	0.91	0.43
2W0X1	0.15	0.15	0.92	0.45
2A6X4	0.16	0.11	0.92	0.43
4A0X1	0.19	0.14	0.93	0.46
1C1X1	0.11	0.10	0.93	0.35

Table A.3—continued

	Average Impacts			
AFSC	EPRs	Decorations	Testing	Longevity
2E2X1	0.12	0.11	0.93	0.36
3E6X1	0.07	0.17	0.94	0.55
3P0X1	0.11	0.12	0.95	0.38
2A6X5	0.13	0.14	0.95	0.43
2S0X2	0.10	0.12	0.95	0.33
2T3X0	0.19	0.16	0.95	0.41
1N2X1	0.09	0.16	0.96	0.36
1A2X1	0.06	0.17	0.96	0.23
1A3X1	0.10	0.18	0.96	0.35
2E1X2	0.09	0.13	0.96	0.43
1C4X1	0.18	0.17	0.97	0.43
3E5X1	0.08	0.15	0.97	0.46
3V0X1	0.14	0.18	0.98	0.49
3E1X1	0.08	0.13	0.98	0.47
3E2X1	0.12	0.10	1.01	0.43
2R0X1	0.17	0.15	1.02	0.46
4N1X1	0.27	0.13	1.02	0.60
1A0X1	0.07	0.16	1.03	0.33
3C1X1	0.06	0.11	1.04	0.37
2A7X4	0.23	0.12	1.07	0.38
2R1X1	0.17	0.12	1.07	0.50
4A2X1	0.13	0.10	1.09	0.41
3E0X1	0.09	0.13	1.14	0.40

APPENDIX B

AFSC Titles

Table B.1 provides titles for AFSCs listed in the tables in Appendix A. Note that some of them are now obsolete.

Table B.1
AFSC Titles

AFSC	Title
1A0X1	In-Flight Refueling
1A1X1	Flight Engineer
1A2X1	Loadmaster
1A3X1	Airborne Mission Systems
1A4X1	Airborne Battle Management
1A5X1	Airborne Missions Systems
1A6X1	Flight Attendant
1A7X1	Aerial Gunner
1A8X1	Airborne Cryptologic Linguist
1C0X2	Operations Resource Management
1C1X1	Air Traffic Control
1C2X1	Combat Control
1C3X1	Command Post
1C4X1	Tactical Air Command and Control
1C5X1	Aerospace Control and Warning Systems
1C6X1	Space Systems Operations
1C7X1	Airfield Management
1N0X1	Intelligence Applications
1N1X1	Imagery Analysis
1N2X1	Communications Signals Intelligence
1N3X0	Cryptologic Linguist (Superintendent)
1N3X1	Germanic Cryptologic Linguist
1N3X2	Romance Cryptologic Linguist
1N3X3	Slavic Cryptologic Linguist
1N3X4	Far East Cryptologic Linguist
1N3X5	Mid-East Cryptologic Linguist

Table B.1—continued

AFSC	Title
1N3X6	African Cryptologic Linguist
1N3X7	Turkic Cryptologic Linguist
1N3X8	Polynesian Cryptologic Linguist
1N3X9	Indo-Iranian Cryptologic Linguist
1N4X1	Network Intelligence Analysis
1N5X1	Electronic Signals Intelligence Exploitation
1N6X1	Electronic System Security Assessment
1S0X1	Safety
1T0X1	Survival, Evasion, Resistance, and Escape
1T1X1	Aircrew Life Support
1T2X1	Pararescue
1W0X1	Weather
1W0X1A	Weather Forecaster
2A0X1	Avionics Test Station and Components
2A0X1A	Avionics Sys, F-15
2A0X1B	Avionics Sys, Helicopters & Aircraft (except F-15)
2A3X1	A-10, F-15, & U-2 Avionics Systems
2A3X1A	Attack Con
2A3X1B	Instm & Flt Con
2A3X1C	Comm, Nav, & Pen Aids
2A3X2	F-16, F-117, RQ-1, CV-22 Avionic Systems
2A3X3	Tactical Aircraft Maintenance
2A3X3A	F-15
2A3X3B	F-16/F-117
2A3X3J	General (except F-15/F-16/F-117)
2A5X1	Aerospace Maintenance
2A5X1J	C-5/C-9/C-12/C-17/C-20/C-21/C-22/C-26/C-27/C-130/ C-141/T-39/T-43
2A5X1K	B-1/B-2/B-52
2A5X1L	C-135/C-18/E-3/E-4/KC10/VC25/VC137
2A5X2	Helicopter Maintenance
2A5X3	Integrated Avionics Systems
2A5X3A	Comm, Nav, & Mission
2A6X1	Aerospace Propulsion
2A6X1A	Jet Engines
2A6X1B	Turboprop & Turboshaft
2A6X2	Aerospace Ground Equipment
2A6X3	Aircrew Egress Systems
2A6X4	Aircraft Fuel Systems
2A6X5	Aircraft Hydraulic Systems
2A6X6	Aircraft Electrical and Environmental Systems
2A7X1	Aircraft Metals Technology
2A7X2	Nondestructive Inspection
2A7X3	Aircraft Structural Maintenance
2A7X4	Survival Equipment
2E0X1	Ground Radar Systems

Table B.1—continued

AFSC	Title
2E1X1	Satellite, Wideband, and Telemetry Systems
2E1X2	Meteorological and Navigation Systems
2E1X3	Ground Radio Communications
2E1X4	Visual Imagery and Intrusion Detection Systems
2E2X1	Com, Network, Switching & Crypto Systems
2E6X2	Communication Cable and Antenna Systems
2E6X3	Voice Network Systems
2F0X1	Fuels
2G0X1	Logistics Plans
2M0X1	Missile and Space Systems Elect Maintenance
2M0X2	Missile and Space Systems Maintenance
2M0X3	Missile and Space Facilities
2P0X1	Precision Measurement Equipment Laboratory
2R0X1	Maintenance Management Analyst
2R1X1	Maintenance Production
2S0X1	Materiel Management
2S0X2	Supply Systems Analysis
2T0X1	Traffic Management
2T1X1	Vehicle Operations
2T2X1	Air Transportation
2T3X0	Vehicle & Vehicular Equip Maintenance (Advanced Level)
2T3X1	Vehicle and Vehicular Equipment Maintenance
2T3X2	Special Vehicle Maintenance
2T3X5	Vehicle Body Maintenance
2T3X7	Vehicle Management & Analysis
2W0X1	Munitions Systems
2W1X1	Aircraft Armament Systems
2W2X1	Nuclear Weapons
3A0X1	Information Management
3C0X1	Communication-Computer Systems Operations
3C0X2	Communication-Computer Systems Programming
3C1X1	Radio Communications Systems
3C1X2	Electromagnetic Spectrum Management
3C2X1	Communication-Computer Systems Control
3C3X1	Comm-Comp Sys Planning & Implementation
3E0X1	Electrical
3E0X2	Electrical Power Production
3E1X1	Heating, Ventilation, AC, & Refrigeration
3E2X1	Pavement and Construction Equipment
3E3X1	Structural
3E4X1	Utilities Systems
3E4X2	Liquid Fuel Systems Maintenance
3E4X3	Pest Management
3E5X1	Engineering
3E6X1	Operations Management
3E7X1	Fire Protection

Table B.1—continued

AFSC	Title
3E8X1	Explosive Ordnance Disposal
3E9X1	Readiness
3H0X1	Historian
3M0X1	Services
3N0X1	Public Affairs
3N0X2	Radio and Television Broadcasting
3N1X1	Regional Band
3N2X1	Premier Band
3P0X1	Security Forces
3S0X1	Personnel
3S1X1	Military Equal Opportunity
3S2X1	Education and Training
3S3X1	Manpower
3V0X1	Visual Information
3V0X2	Still Photographic
3V0X3	Visual Information Production—Documentation
4A0X1	Health Services Management
4A1X1	Medical Materiel
4A2X1	Biomedical Equipment
4B0X1	Bioenvironmental Engineering
4C0X1	Mental Health Service
4D0X1	Diet Therapy
4E0X1	Public Health
4H0X1	Cardiopulmonary Laboratory
4J0X2	Physical Medicine
4M0X1	Aerospace Physiology
4N0X1	Medical Service
4N1X1	Surgical Service
4P0X1	Pharmacy
4R0X1	Diagnostic Imaging
4T0X1	Medical Laboratory
4T0X2	Histopathology
4T0X3	Cytotechnology
4U0X1	Orthotic
4V0X1	Optometry
4Y0X1	Dental Assistant
4Y0X2	Dental Laboratory
5J0X1	Paralegal
5R0X1	Chaplain Assistant
6C0X1	Contracting
6F0X1	Financial Management & Comptroller
7S0X1	Special Investigations

APPENDIX C

WAPS Changes over Time

Table C.1 outlines the initial WAPS formula as implemented for selection to E4–E7 in 1970. Selections to E8 and E9 did not initially fall under WAPS.

Table C.1
WAPS as Implemented on January 2, 1970

Factor	Maximum Points	Calculation
Time in service	40	2 points per completed year of service; 1 point for additional service less than six months; 2 points for additional service greater than six months
Time in grade	60	One-half point per completed month in grade
Decorations	25	Medal of Honor, 15 points AF Cross, 9 points Distinguished Service Cross, 9 points Distinguished Service Medal, 9 points Silver Star, 7 points Legion of Merit, 7 points Airman's Medal, 5 points Soldier's Medal, 5 points Bronze Star, 5 points Meritorious Service Medal, 5 points Air Medal, 3 points Commendation Medal, 3 points Purple Heart, 1 point
Airman Performance Reports (APRs)	135	15 x (average APR score over past five years) or 10 APRs, whichever came first; note that APR scores ranged from 0 to 9
SKT	95	An individual's rank order score
PFE	95	An individual's rank order score

SOURCE: Hall and Nelsen, 1980, pp.53–55.

When the Air Force instituted WAPS, airmen competed for selection within their AFSCs as they do today; however, at that time, the Air Force promoted to fill vacancies. Thus, each AFSC had a different selection rate. Of special interest is the fact that both the PFE and SKT scores were determined by rank-order percentile on the test (in 1-percent increments), not the percentage of correct answers, as is the case today. This meant that testing initially had greater and more uniform impacts across AFSCs than it does today.

Table C.2 outlines eight changes to WAPS since its inception in 1970.

Table C.2
Major Changes to WAPS

Year	Modification
1971	Selection to E4 removed from WAPS
1972	Maximum PFE and SKT points increased to 100 based on percentage of correct answers
1977	E8 and E9 selections fall under WAPS with the following factors: Supervisory Exam, 100 points EPRs, 135 points Professional Military Education (PME), 35 points TIG, 60 points TIS (1/12 point/month up to 25 years), 25 points Decorations, 25 points Board score, 270–450 points
1978	SKT-exempt concept introduced
1985	More points for Purple Heart
1989	Senior NCO Academy points removed for selections to E8 and E9
1991	Greater weight given to more-recent EPRs
1997	PFE scores doubled for SKT-exempt members

SOURCES: Shore and Gould, 2004, pp. A4–A5, B9–B10; Hall and Nelsen, 1980, p. 85.

A Related Policy Change

Starting in 2003, the Air Force implemented an enhanced CCS program to distribute E8 and E9 selections.[1] This represented a substantial deviation from the previous ESO/CCS policy, in part because the Air Force now promotes individuals in some "donor" AFSCs below the baseline rate and promotes others above the standard enhanced rate under CCS (1.2 times the baseline rate). Figure C.1 shows the range of selection rates for the 05E8 cycle for AFSCs with at least 25 eligible E7s.

AF/A1PPP compiles the CCS and donor list with recommendations from various offices that manage the enlisted force. Criteria include projected manning through the selection cycle, pending mergers and conversions, the overall health of career field manning, and the number of those eligible for selection.

Figure C.1
Distribution of Selection Rates for 05E8 Cycle

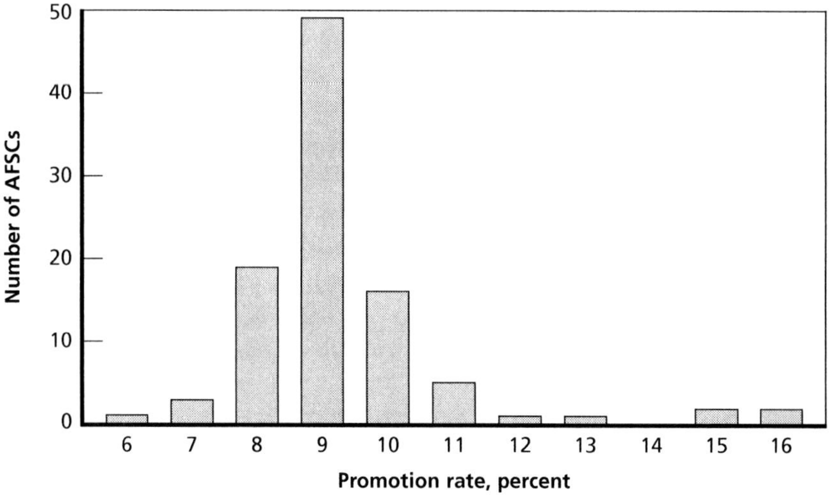

SOURCE: Derived from WAPS history file.
RAND *MG678-C.1*

[1] Voegtle, 2006, p. 1.

Figure C.2 shows the relationship between the 05E8 selection rates and 05E8 manning. While there was some relationship between manning and selection rates, there were obvious exceptions.

The Air Force felt that it was necessary to implement the enhanced CCS program because senior NCO manning for many AFSCs varies widely from 100 percent, as evidenced by an examination of the manning percentages along the horizontal axis in Figure C.2. The Air Force may not realize that the gross deviations from 100 percent manning in Figure C.2 are partially the result of its ESO policy and of not standardizing test scores and not centrally constraining requirements. Hence, the enhanced CCS program does not address root causes.

Figure C.2
05E8 Selection Rates Versus September 30, 2005 Manning

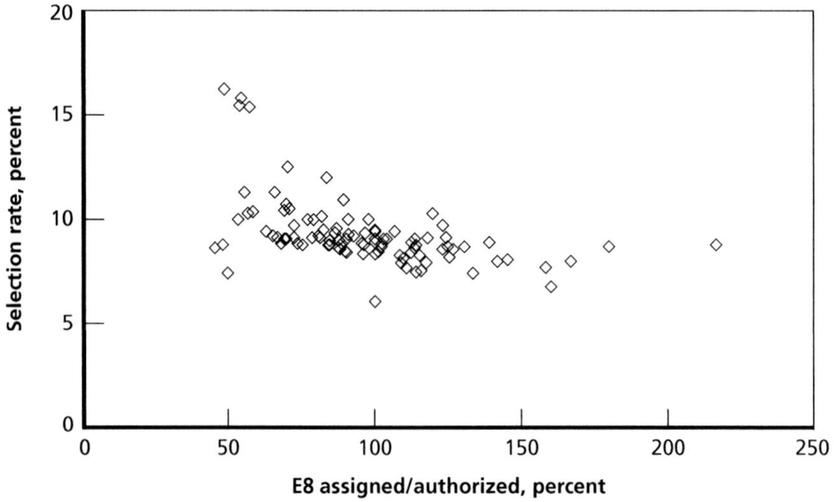

SOURCE: Derived from WAPS history file and RAW.
RAND *MG678-C.2*

APPENDIX D

Periodic WAPS Revalidation

> WAPS has undergone revalidation in 1972, 1977, and 1986. Using a policy-capturing approach for each revalidation, the data indicated that different promotion formulae should be applied to those in different grades. However, it was determined that different weighting schemes would lead to a less understandable, and potentially less acceptable, system.[1]

The above passage indicates that selection outcomes to some grades would differ if the Air Force used selection boards rather than WAPS. Therefore, revalidation does not imply reaffirmation. A 2004 study based on the judgments of 37 E9s also concluded in the following recommendations that the Air Force should modify WAPS weights:[2]

E5–E7 Cycles

1. Greater weight should be given the EPR score for all grades. With the high level of acceptance that WAPS currently enjoys, a small change would probably be most advisable. The alternative is a performance score which has more variance, possibly one used for promotion purposes only.
2. Less weight should be given TIS and TIG. This may be best accomplished with a slight lowering of the caps.

[1] Duncan, 1994, p. 2.

[2] These recommendations are from Shore and Gould, 2004, p. 35.

3. For E5, greater weight should be given to the SKT than to the PFE, but the total for the two tests should stay at the present level. The weight for the SKT should be about twice that given to the PFE. In addition to being the panel's policy, it seems useful to have the more junior grade focus on specialty knowledge. For E6, the technical finding is at the boundary of making no change and having the SKT with about 20% more weight than the PFE. If the latter change is made, there would be a trend through the three WAPS grades with E7 giving them equal weight.

4. Overall, any adjustments to the weightings should be small, even though the panel's policy represented large changes. Changes to a successful program are probably best made on an evolutionary, small scale.

E8–E9 Cycles

1. The Board weight should be increased slightly to accommodate the panel's policy.
2. If the same weights are to be applied to both E8 and E9, no other changes are recommended. If different weights for E8 and E9 are considered, then lowering the caps for longevity (TIS and TIG) for E8 is the next strongest recommendation.

APPENDIX E

Four Approaches to Measuring the Impacts of WAPS Factors

Approach One

One approach for measuring impacts would be to use the standard deviations of WAPS factors for each AFSC.[1] However, this approach has a shortcoming. Figure E.1 shows the distribution of testing scores for two AFSCs for the 05E7 cycle with identical standard deviations, $\sigma = 17.5$.

Even though both AFSCs had the same standard deviation in test scores, we cannot conclude that testing had the same impact in both AFSCs.[2] Table E.1 shows that the standard deviations for the other WAPS components and for the total scores were not equal for AFSCs 2A5X1 and 2E0X1 on the 05E7 cycle. In this case, testing accounted for a greater percentage of the total variation in AFSC 2E0X1 than it did in AFSC 2A5X1.[3] Hence, even though the two AFSCs had the

[1] *Standard deviation* is a common statistical measure of variation.

[2] We selected two AFSCs with the same standard deviations to make a point. In general, there is a wide range of standard deviations of test scores (for example, see Figures 2.16–2.18).

[3] If the four components for an AFSC in Table E.1 were independent, the total variation would exceed the variation of each individual component. In AFSC 2E0X1, the variation in testing could have exceeded the total variation only if its test scores were correlated with some of its other WAPS component scores. In general, there is a negative correlation between test scores and longevity.

Figure E.1
Distribution of Testing Scores in Cycle 05E7 for AFSCs 2A5X1 and 2E0X1

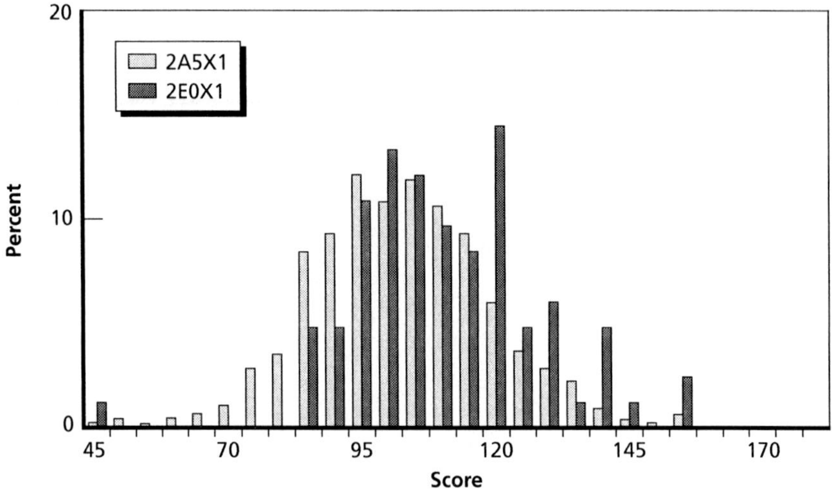

SOURCE: Derived from AFPC WAPS history file.
NOTE: Standard deviation (σ) = 17.5
RAND *MG678-E.1*

Table E.1
Standard Deviations of WAPS Components in Cycle 05E7 for AFSCs 2A5X1 and 2E0X1

AFSC/Cycle	Testing	Longevity	Decorations	EPRs	Total
2A5X1/05E7	17.5	9.2	4.3	3.9	18.8
2E0X1/05E7	17.5	9.4	3.6	4.3	17.4

SOURCE: Derived from WAPS history file.

same variance in their test scores, testing played a greater role in the variation of the total score in AFSC 2E0X1.

Approach Two

A second approach would be to divide the standard deviation of each AFSC's WAPS factors by the standard deviation of its total score. These ratios would normally range between 0 and 1. Table E.2 illustrates this approach for our example AFSCs. As we would expect, Approach Two indicates that 2E0X1 had a greater testing impact.

Although Approach Two would take the variation of other scores into account, it would not account for correlations that often occur between WAPS components. Tables E.3 and E.4 illustrate these correlations for our example AFSCs. For example, in AFSC 2E0X1, the negative EPR/Longevity correlation coefficient of –.37 indicates that those with higher EPR scores tended to have lower longevity scores, as we would expect. Similarly, those with higher longevity scores tended to have lower testing scores. We see the same relationships in AFSC 2A5X1, but they are not as strong.[4]

Table E.2
Standard Deviations of WAPS Components Divided by Standard Deviation of Total Scores in Cycle 05E7 for AFSCs 2A5X1 and 2E0X1

AFSC/Cycle	Testing	Longevity	Decorations	EPRs	Total
2A5X1/05E7	0.93	0.49	0.23	0.21	1.0
2E0X1/05E7	1.01	0.54	0.20	0.25	1.0

SOURCE: Derived from WAPS history file.

Table E.3
Correlation Matrix in Cycle 05E7 for AFSC 2E0X1

	EPR	Decorations	Testing	Longevity
EPR	1.00			
Decorations	.13	1.00		
Testing	.33	– 0.10	1.00	
Longevity	–.37	0.13	–0.44	1.00

SOURCE: Derived from WAPS history file.

[4] Correlation coefficients closer to 1 or –1 indicate stronger positive or negative relationships. Coefficients closer to 0 indicate weaker relationships.

Table E.4
Correlation Matrix in Cycle 05E7 for AFSC 2A5X1

	EPR	Decorations	Testing	Longevity
EPR	1.00			
Decorations	0.28	1.00		
Testing	0.10	−0.05	1.00	
Longevity	−0.19	0.11	−0.26	1.00

SOURCE: Derived from WAPS history file.

We show later that there is a cause-and-effect relationship between testing and longevity. There can be other cause-and-effect relationships. For example, some individuals who do not believe they possess sufficient EPR points to be competitive in a selection cycle may not exert their best efforts on tests.[5]

Approach Three

This approach measures the degree to which each individual's rank-ordered percentile changes as we remove a WAPS component from the equation. We used this technique when we introduced the concept of variation in Chapter Two. This approach is intuitive and accounts for each individual's correlated component scores. For example, if an individual with low longevity points elected not to study for the SKT, that person's rank order would not change as dramatically when calculated with and without test scores. Approach Three also accounts for the differential variation in other factors. Table E.5 illustrates Approach Three for our example AFSCs. In AFSC 2E0X1, the average individual's rank order changed by 28.0 percentile points (up or down) when we recalculated his/her WAPS score without using test scores. Again, Approach Three indicates that testing has the highest impact in both AFSCs, with the greater impact in AFSC 2E0X1.

---

[5] The Air Force does not completely rewrite SKTs every year, and Air Force members are not permitted to study together for promotion purposes. Hence, individuals sometimes take the PFE or SKT in anticipation of seeing some of the same questions when they test the following year with more longevity points.

Table E.5
Approach Three: Average Change in Rank Order Percentile

AFSC/Cycle	Minus Testing	Minus Longevity	Minus Decorations	Minus EPRs
2A5X1/05E7	24.2	11.9	5.2	3.0
2E0X1/05E7	28.0	13.5	5.4	3.0

SOURCE: Derived from WAPS history file.

However, using rank order implicitly assumes that WAPS scores are uniformly distributed when in fact they tend to be normally distributed.[6] This assumption could tend to overstate the impact of WAPS factors for individuals near the densely populated means and understate impacts for airmen in the sparse tails of the WAPS score distributions.

Approach Four

We ultimately adopted a fourth approach, which has the best properties of the approaches we considered. It calculates for each tester the absolute difference in the number of standard deviations that he/she is away from the mean, with and without each WAPS component score. The statistic for each WAPS component is the average of the absolute changes in standard deviations for each member of each AFSC. This approach accounts for correlated component scores and does not transform distributions. It also accounts for the variation in other factors. Table E.6 shows statistics using Approach Four for our example AFSCs. For example, individuals in AFSC 2E0X1 moved an average of 0.943 standard deviations when we removed testing from their total WAPS scores.

---

6 A *uniform distribution* would have about the same number of observations for each possible WAPS score. A *normal distribution* is the classic bell-shaped curve with significantly more occurrences of some WAPS scores than others.

Table E.6
Approach Four: Average Change in Standard Deviations from the Mean

AFSC/Cycle	Minus Testing	Minus Longevity	Minus Decorations	Minus EPRs
2A5X1/05E7	0.839	0.391	0.182	0.117
2E0X1/05E7	0.943	0.415	0.171	0.143

SOURCE: Derived from WAPS history file.

As an excursion to Approach Four, we eliminated individuals with extremely low testing scores (worse than random guessing). Extremely low scores inflate the impact of testing. For example, we occasionally observed individuals who scored 0 on both the PFE and SKT. Random guessing would have produced scores of about 25 on each test, so we conclude that these individuals simply wrote their names on the tests and departed.[7] There are a number of rational reasons that an individual might elect to this.

- Some do not seek selection because assuming a higher grade might make them more vulnerable for duties, assignments, or responsibilities they do not wish to accept.
- Some may not believe that even their best testing effort would get them promoted, but they want exposure to the test questions to help them prepare for future selection cycles.
- Taking the exam ensures that an individual remains eligible for selection. Because the total number of selections an AFSC receives is a function of the number of eligibles, taking the test might create one more selection for the AFSC.

Ultimately, we elected not to eliminate the few individuals with extremely low test scores because including them had no practical impact on our ultimate objective.

---

7 The Air Force expects all eligible members to be tested. Members must provide explanations to their supervisors when they miss scheduled testing appointments.

APPENDIX F

Multivariate Models to Predict Selection Rates to E5

This appendix describes the models that we developed to predict selection rates to E5 as a function of time in service. It also provides insight into the dynamics of the predictor variables.

Each year, AF/A1 (Deputy Chief of Staff for Manpower and Personnel) strength managers establish the selection rates to grades E5 through E9. After accounting for current strengths by grade and projected losses by grade, these rates should allow force managers to satisfy funded aggregate strengths by grade. Figure F.1 shows that managers used a wide range of selection rates over the 98E5–05E5 cycles to control E5 strength. As we analyzed the WAPS factors that drove selection outcomes, we were mindful of these selection rates.

In addition to accounting for the wide variation in selection rates for the 98E5–05E5 cycles, we also considered the selection rate differentials between "healthy" AFSCs and those with CCS skills (Figure F.2).[1]

The high selection rates for 00E5–03E5 also modified the experience levels of E4s. Table F.1 shows the TIS distributions of the E4s competing in the 98E5–05E5 cycles. Because of fixed phase points to E4, the leading edge of the eligibles, from a longevity perspective, remained stable at four years TIS. However, the gradual compression

[1] The E5–E7 selection cycle rate for CCS skills is set at 1.2 times the rate for healthy AFSCs. AFSC/grade combinations move on and off of the CCS list as their manning pictures change.

Figure F.1
E5 Cycle Selection Rates

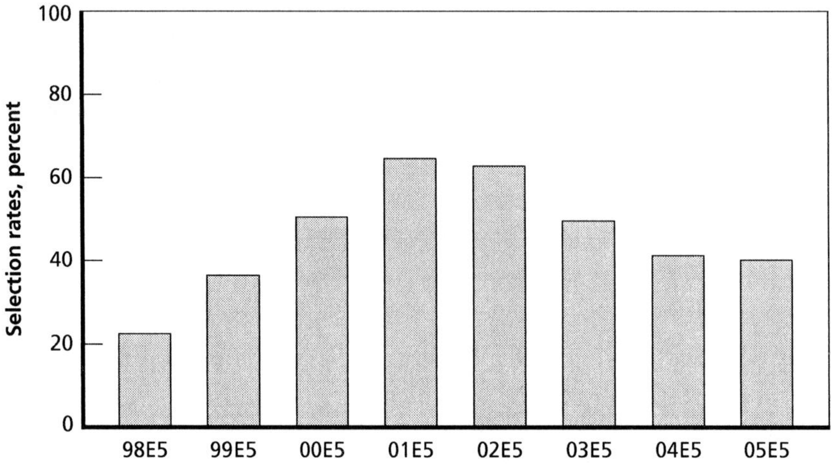

SOURCE: AFPC, Directorate of Personnel Programs (AFPC/DPP).
RAND *MG678-F.1*

Figure F.2
Normal and CCS Selection Rates, 98E5–05E5 Cycles

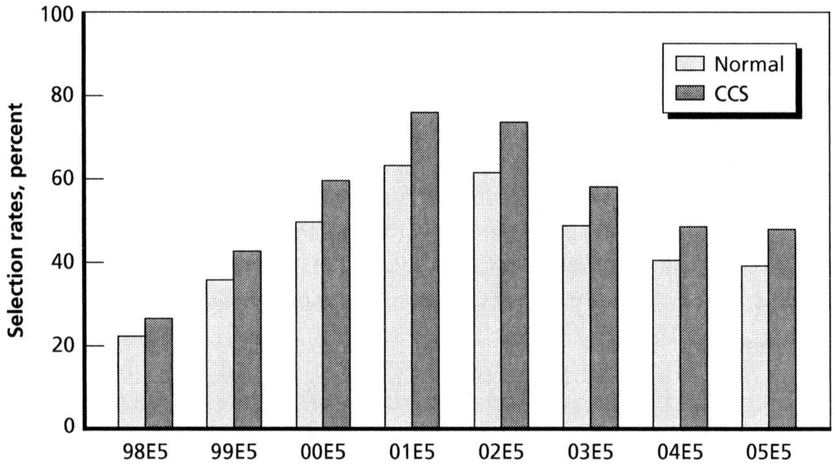

SOURCE: WAPS history file.
RAND *MG678-F.2*

of E4s toward the left side of Table F.1[2] reduced the variation (standard deviation) of longevity scores (Figure F.3). In turn, this reduced the impact of longevity (Figure F.4).

Figure F.4 also demonstrates that the impact of EPRs increased between 1998 and 2003. Even though supervisors were awarding higher percentages of perfect EPRs (Figure F.5), the relative impact of EPRs increased because the variation in longevity scores was decreasing at an extreme rate. The impact of EPRs also increased initially because higher percentages of perfect scores actually increased standard deviations (see Figure F.3) when only 40 percent had perfect EPR scores. Hence, the standard deviation and impact of EPR scores increased through the 03E5 cycle.

Table F.2 lists our candidate independent (predictor) variables. These variables were candidates because we hypothesized that they might have affected selection rates. The models also contain indicator variables to account for differences in selection rates across the eight selection cycles spanned by the underlying data.

Table F.1
Eligible E4s by Time in Service (%)

	Time in Service (years, at end of cycle)							
Cycle	**4**	**5**	**6**	**7**	**8**	**9**	**10**	**Total**
98E5	6	24	21	21	13	8	7	100
99E5	8	26	21	17	14	7	6	100
00E5	10	34	22	16	9	5	3	100
01E5	14	41	23	12	6	2	2	100
02E5	22	46	19	8	3	1	1	100
03E5	28	42	18	7	3	1	1	100
04E5	23	46	18	8	3	1	1	100
05E5	24	39	23	8	3	1	1	100

SOURCE: WAPS history file.

[2] This compression was driven primarily by the interaction of high selection rates, funded E5 authorizations, and small year groups (the AF has consistently reduced accessions to reduce end strengths).

Figure F.3
Trends in EPR and Longevity Standard Deviations, 98E5–05E5 Cycles

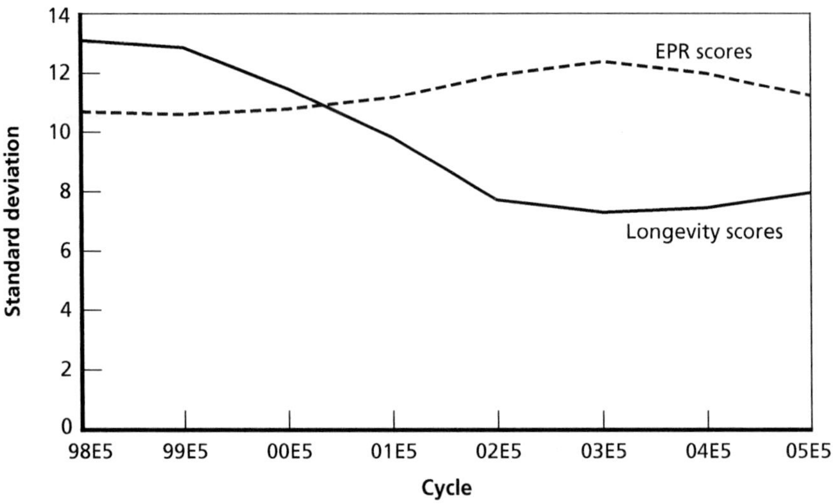

SOURCE: Derived from WAPS history file.
RAND *MG678-F.3*

Figure F.4
WAPS Factor Impacts, 98E5–05E5 Cycles

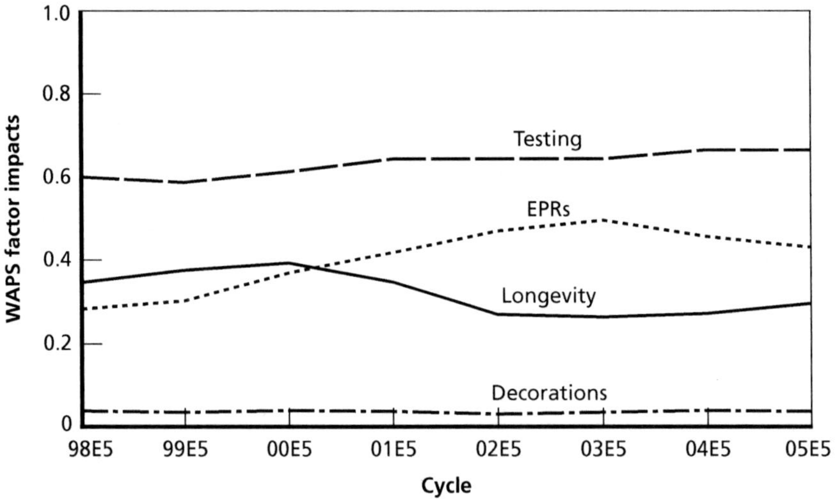

SOURCE: Derived from WAPS history file.
RAND *MG678-F.4*

Figure F.5
Perfect EPR Scores, 98E5–05E5 Cycles

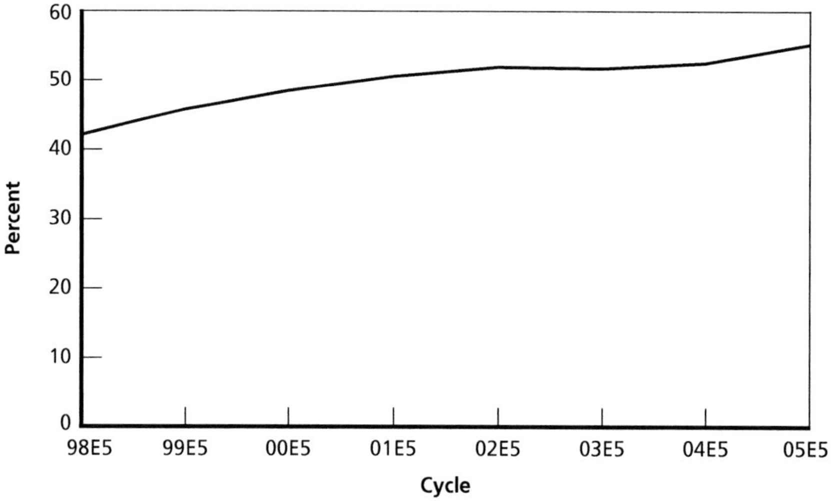

SOURCE: Derived from WAPS history file.
RAND *MG678-F.5*

Table F.2
Candidate Predictor Variables

Predictor Variables
EPR impact for the AFSC/cycle (Approach Four)
Decoration impact for the AFSC/cycle (Approach Four)
Testing impact for the AFSC/cycle (Approach Four)
Longevity impact for the AFSC/cycle (Approach Four)
98E5 indicator variable (0=no or 1=yes)
99E5 indicator variable (0 or 1)
00E5 indicator variable (0 or 1)
01E5 indicator variable (0 or 1)
02E5 indicator variable (0 or 1)
03E5 indicator variable (0 or 1)
04E5 indicator variable (0 or 1)
CCS indicator variable for the AFSC/cycle (0 or 1)

We did not use all AFSCs in our models. In this and similar analyses for selections to grades E6 and E7, we eliminated all tax AFSCs. We also excluded AFSCs that had widely varying testing impacts across the 98E5 to 05E5 cycles. We eliminated these unstable AFSCs because their interactions between longevity and testing were complex and dynamic across selection cycles. Retaining unstable AFSCs would have made it much more difficult to isolate the relationships we were trying to observe. Our exclusions left us with the 87 AFSCs listed in Appendix A, Table A.1. These AFSCs across eight selection cycles offered the possibility of 696 (87 AFSCs times 8 cycles) observations per model. However, we developed our model for those with four years TIS based on the 385 occasions when stable AFSCs had at least 25 E4 eligibles with four years TIS.[3]

To understand if predictor variables had the same impacts on selection rates for junior and senior individuals within AFSCs, we developed four different models based on seniority. For E5 selection cycles, we modeled the four groups shown in Table F.3.

Table F.4 lists our TIS = 4 model results. The significant predictor variables have the indicated coefficients, and the variables that are not significant have missing coefficients. The adjusted R^2 is 0.88, which is exceptional for a model that predicts human behavior.[4] The magni-

Table F.3
E5 Cycle Models

Model	Seniority Group
TIS = 4	Four years TIS
TIS = 5	Five years TIS
TIS = 6	Six years TIS
TIS = 7	Seven years TIS

[3] The 385 occasions captured 32,681 enlisted members with four years of service, which was 73 percent of the total before any exclusions.

[4] R^2 is a measure of how much of the variability in selection rates is explained using this collection of variables and coefficients in this linear model. A model that perfectly explained the data would have $R^2 = 1$. A model with no explanatory power would have $R^2 = 0$.

Table F.4
TIS = 4 Model

Predictor Variable	Coefficients	P-values
EPR impact	0.49	0.000000
Decoration impact		
Testing impact	0.50	0.000000
Longevity impact	−0.63	0.000000
98E5 indicator variable (0 or 1)	−0.15	0.000000
99E5 indicator variable (0 or 1)	−0.07	0.000022
00E5 indicator variable (0 or 1)	0.14	0.000000
01E5 indicator variable (0 or 1)	0.30	0.000000
02E5 indicator variable (0 or 1)	0.24	0.000000
03E5 indicator variable (0 or 1)	0.09	0.000000
04E5 indicator variable (0 or 1)		
CCS indicator variable (0 or 1)	0.06	0.000010
Intercept[a]		

[a] If none of the other predictor variables were explanatory, the value of the intercept would be the average of the AFSC selection rates for those with four years TIS. When a model has predictive power, intercepts have values that are different from the average of the dependent variable.

tudes of the P-values indicate that the listed coefficients are statistically significant.[5] The coefficients in Table F.4 are intuitive. For example, we would expect higher selection rates for CCS AFSCs. For those with four years TIS, our model adds 0.06 (six percentage points) to the selection rates for CCS AFSCs. The model also adds a factor ranging from −0.15 to 0.3 to account for differences among the selection cycles represented in the data.

To capture the effects of different EPR impacts across AFSCs, our TIS = 4 model adds the product of 0.49 and the AFSC's EPR impact for that cycle as measured using Approach Four in Chapter Two. Because the coefficient (0.49) is positive, we can deduce that the

[5] The *P-value* is the probability of concluding that a predictor variable is statistically significant when it really is not. For our models, we retained predictor variables with P-values of less than 5 percent (0.05).

Air Force promoted E4s with four years TIS at higher rates when they were in AFSCs that had greater EPR impacts. We saw in Figure 2.7 that EPR impacts varied from 0.2 to 0.6 for the 05E5 cycle. Multiplying these impacts by our EPR coefficient means that, all other things being equal, the Air Force promoted individuals with four years TIS in AFSCs with the highest EPR impacts about (0.49 ×0.6) – (0.49 × 0.2) = 0.20, or 20 percentage points higher than in AFSCs with the lowest EPR impacts.

In the TIS = 4 model, the longevity coefficient is negative. This means that the Air Force promoted individuals with four years TIS at lower rates when they were in AFSCs with greater variations in longevity scores. Again, this is reasonable. Junior E4s with four years TIS did not compete as well when they faced higher concentrations of individuals with more longevity points, all other things being equal. Figure 2.7 shows that the range of longevity impacts for the 05E5 cycle was 0.2 to 0.4. Individuals with four years TIS in AFSCs with the highest longevity impacts lost about (–0.63 × 0.4) – (–0.63 × 0.2) = –0.13 or 13 percentage points compared to those in AFSCs with the lowest longevity impacts.

Because the coefficient for the decoration impact was not statistically different from zero, we did not include it in the model. Excluding decorations, especially for those with four years TIS, is consistent with Figure 2.7, which shows that decorations had the lowest impact of any WAPS component on the 05E5 cycle.

Finally, the positive coefficient for testing impact indicates that the Air Force was more likely to promote E4s with four years TIS when they were in AFSCs that had higher variations in test scores. This, too, is reasonable because there are higher concentrations of good testers in the four-year TIS groups. Figure 2.7 shows that the range of testing impacts for the 05E5 cycle was 0.5 to 0.9. Therefore, all other things being equal, those with four years TIS in AFSCs that had the highest testing impacts enjoyed about a (0.50 × 0.9) – (0.50 × 0.5) = 0.20 or a 20-percentage-point advantage over those with four years TIS in AFSCs with the lowest testing impacts. Thus, our regression model reveals that the univariate perspective in Figure 4.3 substantially understates the impact of testing differences.[6]

To illustrate the entire calculation, Table F.5 estimates the selection rates for two hypothetical non-CCS AFSCs for the 01E5 cycle that were equal in all respects except testing impact.

Each estimated selection rate is the sum of the products of the model coefficients and the AFSC values. For example, the estimated selection rate for AFSC A:

$$0.61 = (0.49 \times 0.5) + (0.50 \times 0.5) + (-0.63 \times 0.3) + (-0.15 \times 0) + (-0.07 \times 0) + (0.14 \times 0) + (0.30 \times 1) + (0.24 \times 0) + (0.09 \times 0) + (0.06 \times 0).$$

Table F.5
TIS = 4 Modeled Selection rates for High/Low Testing Impacts, Cycle 01E5

Predictor Variable	Coefficients	AFSC A Values	AFSC B Values
EPR impact	0.49	0.50	0.50
Decoration impact			
Testing impact	0.50	0.50	0.90
Longevity impact	−0.63	0.30	0.30
98E5 indicator variable (0 or 1)	−0.15	0	0
99E5 indicator variable (0 or 1)	−0.07	0	0
00E5 indicator variable (0 or 1)	0.14	0	0
01E5 indicator variable (0 or 1)	0.30	1	1
02E5 indicator variable (0 or 1)	0.24	0	0
03E5 indicator variable (0 or 1)	0.09	0	0
04E5 indicator variable (0 or 1)			
CCS indicator variable (0 or 1)	0.06	0	0
Intercept			
Estimated Selection rate		0.61	0.81

[6] For those with four years TIS, Figure 4.3 estimated that E4s in AFSCs with the highest testing impacts had an eight selection point advantage over their counterparts in AFSCs that had the lowest testing impacts.

Table F.6 shows the model for those with TIS = 7. This model has an adjusted R^2 of (0.86), and it used 324 AFSCs/cycles that had at least 25 E4 eligibles with seven years TIS.

Again, the coefficients for this model are intuitive if we realize that E4s competing for E5 with seven years TIS were either poor testers or had less than perfect EPR scores. The negative coefficient for EPR impact means that there was downward pressure on selection rates for individuals with seven years TIS in AFSCs that had more variation in EPR scores.

Similarly, a negative coefficient for testing impact means that selection rates for those with seven years TIS tended to be lower in AFSCs that had more variation in test scores.

Figure F.6 plots the coefficients for testing and EPR impacts for our four E5 cycle models. Junior E4s who were good testers had a greater advantage when they were in AFSCs with greater variations in testing scores. Conversely, senior E4s were disadvantaged in AFSCs with more variation in testing scores.

Table F.6
TIS = 7 Model

Predictor Variable	Coefficients	P-values
EPR impact	−0.18	0.046139
Decoration impact	−1.34	0.030023
Testing impact	−0.37	0.000000
Longevity impact		
98E5 indicator variable (0 or 1)	−0.33	0.000000
99E5 indicator variable (0 or 1)	−0.11	0.000001
00E5 indicator variable (0 or 1)	0.12	0.000000
01E5 indicator variable (0 or 1)	0.23	0.000000
02E5 indicator variable (0 or 1)	0.18	0.000000
03E5 indicator variable (0 or 1)		
04E5 indicator variable (0 or 1)	−0.05	0.003266
CCS indicator variable (0 or 1)	0.10	0.000005
Intercept	0.90	0.000000

Figure F.6
Model Coefficients for 98–05 E5 Cycles

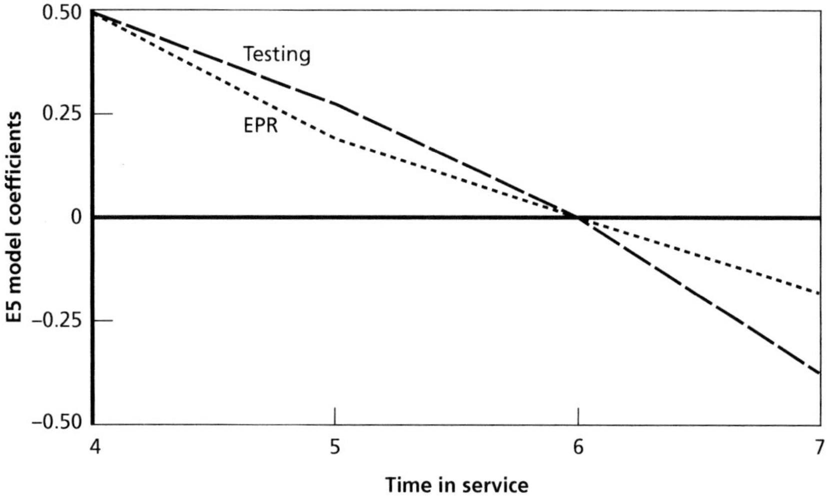

RAND *MG678-F.6*

APPENDIX G

Multivariate Models to Predict Selection Rates to E6

This appendix describes the models that we developed to predict selection rates to E6 as a function of time in service. It also provides insight into the dynamics of the predictor variables.

Figure G.1 shows that managers established a wide range of selection rates over the 98E6–05E6 cycles to control E6 strength.

Figure G.1
E6 Cycle Selection Rates

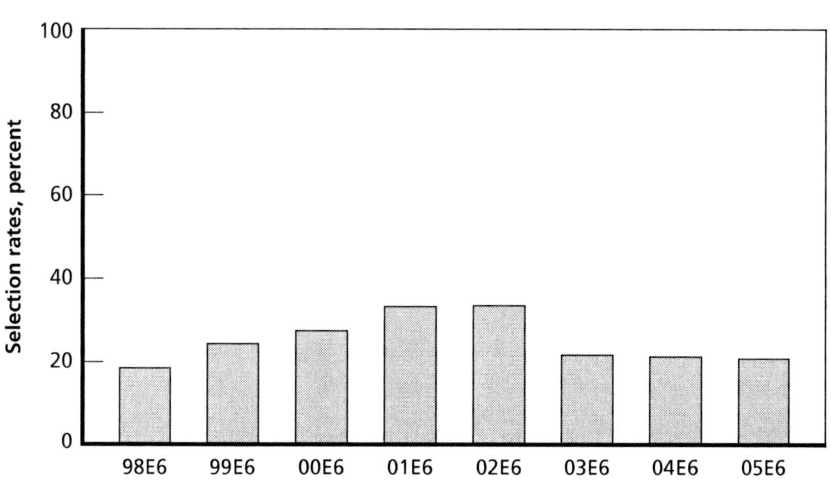

SOURCE: AFPC/DPP.

RAND *MG678-G.1*

Higher selection rates to E5 and to E6 modified seniority patterns. Table G.1 shows the TIS distributions of the E5s competing in the 98E6–05E6 cycles. Between the 98E6 and 05E6 cycles, the E5s competing for selection gradually became more junior.

The shaded cells in Table G.1 indicate the leading edge, by TIS, of the E5s competing for selection changed over the 98E6–05E6 cycles. We used the TIS boundaries defined by these shaded cells to develop our fast burner (FB) model. We employed this fast burner grouping because we found that relative seniority was a better predictor than absolute TIS. For our FB+1+2 model, we considered those with one and two years of service beyond our fast burner groups. Following the same pattern, we developed FB+3+4, FB+5+6, and FB+7+8 models.

Tax AFSC, small AFSC, and other exclusions left us with the 103 AFSCs listed in Appendix A, Table A.2. These 103 AFSCs across eight selection cycles offered the possibility of 824 observations per model. Our FB model actually found 148 AFSCs/cycles that had at least 25

Table G.1
Distribution of E5s by Time in Service (%)

	Time in Service (at end of cycle)												
Cycle	≤7	8	9	10	11	12	13	14	15	16	17	18+	Total
98E6	0	1	3	6	9	13	17	15	12	7	6	12	100
99E6	0	1	3	5	10	11	15	16	13	9	5	12	100
00E6	0	1	4	6	9	14	13	14	14	10	6	9	100
01E6	0	2	4	9	10	13	15	12	11	10	6	7	100
02E6	1	4	8	11	16	13	12	12	7	6	5	5	100
03E6	3	10	13	15	14	15	9	8	6	3	2	2	100
04E6	5	15	16	15	13	11	10	6	4	3	1	2	100
05E6	8	17	18	15	12	10	7	6	3	2	1	1	100

SOURCE: WAPS history file.
NOTE: Shaded cells indicate fast burners.

E5 fast burner eligibles in a single cycle.[1] Table G.2 lists our regression results.

We would have anticipated higher selection rates for CCS AFSCs. However, for the fast burners in the cycles we modeled, the selection advantage was not statistically significant.

Because the EPR impact coefficient (0.33) is positive, we can conclude that the Air Force promoted fast burner E5s at higher rates when they were in AFSCs with greater variations in EPR scores.

The positive coefficient for testing impact indicates that the Air Force promoted fast burner E5s at higher rates when they were in AFSCs that had greater variations in test scores. Figure 2.12 shows that the range of testing impacts for the 05E6 cycle was (0.3) to (0.9). Therefore, all other things being equal, fast burners in AFSCs that had the highest testing impact enjoyed a $(0.49 \times 0.9) - (0.49 \times 0.3) = 0.29$ or a

Table G.2
Fast Burner Model

Predictor Variable	Coefficients	P-values
EPR impact	0.33	0.000080
Decoration impact		
Testing impact	0.49	0.000000
Longevity impact		
98E6 indicator variable (0 or 1)		
99E6 indicator variable (0 or 1)		
00E6 indicator variable (0 or 1)	0.04	0.000008
01E6 indicator variable (0 or 1)	0.06	0.000000
02E6 indicator variable (0 or 1)	0.05	0.000000
03E6 indicator variable (0 or 1)		
04E6 indicator variable (0 or 1)		
CCS indicator variable (0 or 1)		
Intercept	–0.32	0.000000

[1] The 148 data points captured 7,270 fast burners, which was 44 percent of the total before all exclusions. The FB+1+2 and the other three models used higher percentages of the total eligibles because there were more instances with at least 25 eligibles.

29-percentage-point advantage over equally experienced fast burners in AFSCs with the lowest testing impacts. The magnitude of this difference is quite remarkable, and it illustrates that the difference of 10 percentage points that we saw in the univariate perspective presented in Figure 4.10 substantially underestimates testing impact.

The adjusted R^2 for the FB model is 0.53, which is lower than the adjusted R^2 for the E5 models. One explanation for the difference is that the eligibles for the E6 cycles had more variation in their longevity scores. We grouped multiple cells together for our E6 models to capture more AFSCs with at least 25 eligibles in a cycle. However, these groupings were necessarily composed of less-homogenous individuals from a longevity perspective.

Table G.3 shows the model for those with seven or eight years beyond the fast burners. This model has an adjusted R^2 of (0.53), and it used 393 AFSCs/cycles that had at least 25 E5 eligibles in a cycle.

Again, the coefficients for this model are intuitive if we realize that senior E5s competing for E6 tend to be poor testers. The negative

Table G.3
FB+7+8 Model

Predictor Variable	Coefficients	P-values
EPR impact		
Decoration impact		
Testing impact	–0.44	0.000000
Longevity impact	–0.11	0.021587
98E6 indicator variable (0 or 1)	–0.07	0.000000
99E6 indicator variable (0 or 1)		
00E6 indicator variable (0 or 1)	0.05	0.000023
01E6 indicator variable (0 or 1)	0.15	0.000000
02E6 indicator variable (0 or 1)	0.15	0.000000
03E6 indicator variable (0 or 1)		
04E6 indicator variable (0 or 1)		
CCS indicator variable (0 or 1)	0.04	0.000356
Intercept	0.74	0.000000

coefficient for testing impact confirms that selection rates for senior E5s tended to be lower in AFSCs that had more variation in test scores.

Figure G.2 plots the coefficients for testing and EPR impact for our five E6 cycle models. Fast burners enjoyed greater advantages when they were in AFSCs with higher variations in testing and EPR scores. Conversely, senior E5s were disadvantaged in AFSCs with more variation in testing and EPR scores.

Figure G.2
Model Coefficients, E6 Cycles

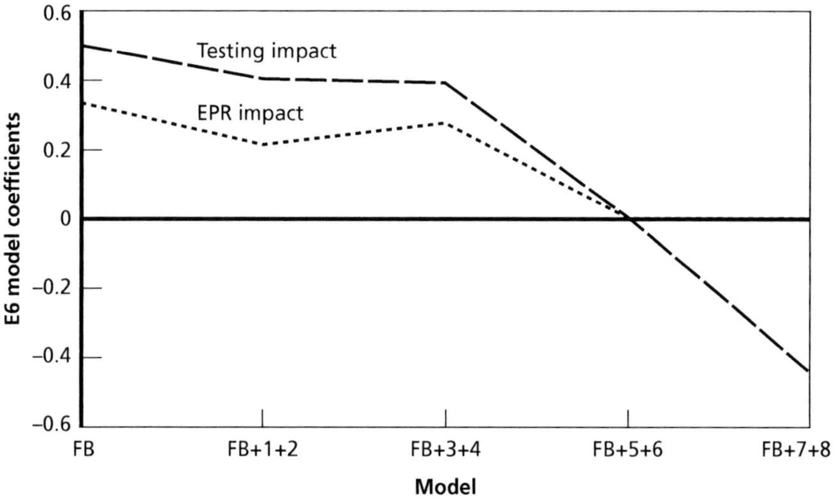

RAND *MG678-G.2*

APPENDIX H

Multivariate Models to Predict Selection Rates to E7

This appendix describes the models that we developed to predict selection rates to E7 as a function of time in service. It also provides insight into the dynamics of the predictor variables.

Figure H.1 shows that strength managers employed a range of selection rates over the 98E7–05E7 cycles.

Figure H.1
E7 Cycle Selection Rates

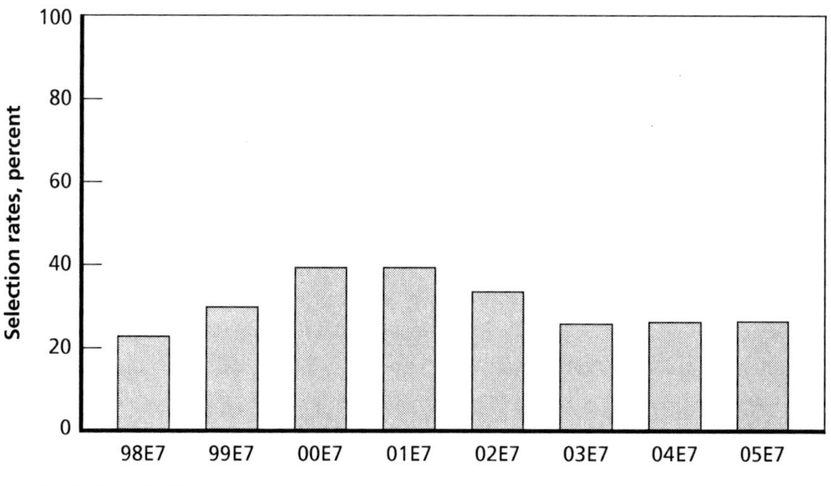

SOURCE: AFPC/DPP.
RAND *MG678-H.1*

Table H.1 shows the TIS distributions of the E6s competing in the 98E6–05E6 cycles. Across the 98E7 and 05E7 cycles, these distributions remained stable for the junior E6s, our population of primary interest.

Unlike with E6 cycles, we did not find it necessary to account for dynamic seniority within fast burners (Table G.1). Exclusions left us with the 84 AFSCs listed in Appendix A, Table A.3. These 84 AFSCs across eight selection cycles offered the possibility of 672 (= 84 × 8) observations per model. Our TIS ≤14 model had 209 AFSCs/cycles that had at least 10 E6 fast burner eligibles in a single cycle.[1]

Table H.2 lists our regression results for the fast burners.

We would have anticipated higher selection rates for CCS AFSCs. However, for the fast burners in the cycles we modeled, the selection advantage was not statistically significant.

Table H.1
Distribution of Eligible E6s by Time in Service (%)

	Time in Service (at end of cycle)										
Cycle	≤11	12	13	14	15	16	17	18	19	20+	Total
98E7	0	1	2	4	9	11	15	19	17	21	100
99E7	0	0	1	4	7	12	14	17	20	25	100
00E7	0	0	1	3	6	11	16	16	17	30	100
01E7	0	0	1	2	5	10	15	19	17	30	100
02E7	0	0	1	3	4	9	15	18	18	31	100
03E7	0	1	1	2	5	8	12	18	19	34	100
04E7	0	1	2	3	5	9	11	15	18	36	100
05E7	1	1	2	5	6	9	13	12	15	36	100

SOURCE: WAPS history file.
NOTE: E6s in shaded cells were fast burners.

[1] Because there were fewer E7s, we reduced the minimum number of eligibles to 10. Using this criterion, we captured 4,848 eligibles with 14 or fewer years of service (52 percent of the total before all exclusions).

Table H.2
TIS ≤14 Model

Predictor Variable	Coefficients	P-values
EPR impact	0.55	0.030185
Decoration impact		
Testing impact	0.43	0.000000
Longevity impact	–0.56	0.000470
98E6 indicator variable (0 or 1)	0.06	0.031749
99E6 indicator variable (0 or 1)	0.13	0.000006
00E6 indicator variable (0 or 1)	0.22	0.000000
01E6 indicator variable (0 or 1)	0.30	0.000000
02E6 indicator variable (0 or 1)	0.15	0.000004
03E6 indicator variable (0 or 1)	0.09	0.003615
04E6 indicator variable (0 or 1)	0.09	0.000199
CCS indicator variable (0 or 1)		
Intercept		

Because the EPR impact coefficient (0.55) in the TIS ≤14 model is positive, we can conclude that the Air Force promoted fast burner E6s at higher rates when they were in AFSCs with greater variations in EPR scores.

We saw in Figure 2.15 that EPR impacts for 05E7 cycle varied from 0 to 0.2. Multiplying these by our EPR coefficient means that, all other things being equal, the Air Force promoted fast burners in AFSCs with the highest EPR score impacts at a rate about 11 percentage points higher than those with the lowest EPR score impacts $[(0.55 \times 0.2) - (0.55 \times 0) = 0.11]$.

The positive coefficient for testing impact indicates that the Air Force promoted E6s with 14 or fewer years TIS at higher rates when they were in AFSCs that had higher variations in test scores. Figure 2.15 shows that the range of testing impacts for the 05E7 cycle was 0.6 to 1.0. Therefore, all other things being equal, fast burners in AFSCs that had the highest testing impact enjoyed about $(0.43 \times 1.0) - (0.43 \times 0.6) = 0.17$ or a 17-percentage-point advantage over equally experienced E6s in AFSCs with the lowest testing impact.

The adjusted R^2 for the TIS ≤14 model is 0.46, which is lower than the adjusted R^2 for the E5 models. One explanation for the difference is that the eligibles for the E7 cycles have more variation in their longevity scores. We grouped multiple longevity cells together for our E7 models to capture more AFSCs with at least 10 eligibles in a cycle. However, larger groupings are necessarily composed of less-homogenous individuals from a longevity score perspective. Also, selection rates based on as few as 10 individuals tend to be more volatile.

Table H.3 shows the model for E6s with ≥19 years TIS. This model has an adjusted R^2 of 0.60, and it used 571 AFSCs/cycles that had at least 10 E5 eligibles in a cycle.

Again, the coefficients for this model are intuitive if we note that senior E6s competing for E7 tend to be poor testers or have less than perfect EPR scores. Negative coefficients for testing and EPR impacts mean that selection rates for senior E6s were lower in AFSCs that had more variation in test and EPR scores.

Table H.3
TIS ≥19 Model

Predictor Variable	Coefficients	P-values
EPR impact	−0.18	0.000373
Decoration impact		
Testing impact	−0.29	0.000000
Longevity impact	0.21	0.000000
98E6 indicator variable (0 or 1)	−0.05	0.000000
99E6 indicator variable (0 or 1)	0.02	0.000156
00E6 indicator variable (0 or 1)	0.09	0.000000
01E6 indicator variable (0 or 1)	0.11	0.000000
02E6 indicator variable (0 or 1)	0.06	0.000000
03E6 indicator variable (0 or 1)		
04E6 indicator variable (0 or 1)		
CCS indicator variable (0 or 1)	0.05	0.000000
Intercept	0.46	0.000000

Figure H.2 plots the coefficients for testing and EPR impact for our four E7 cycle models. As with E4s and E5s, fast burning E6s had greater advantages when they were in AFSCs with higher variations in testing and EPR scores. Conversely, senior E6s were disadvantaged in AFSCs with more variation in test and EPR scores.

Figure H.2
Regression Coefficients, 98E5–05E7 Cycles

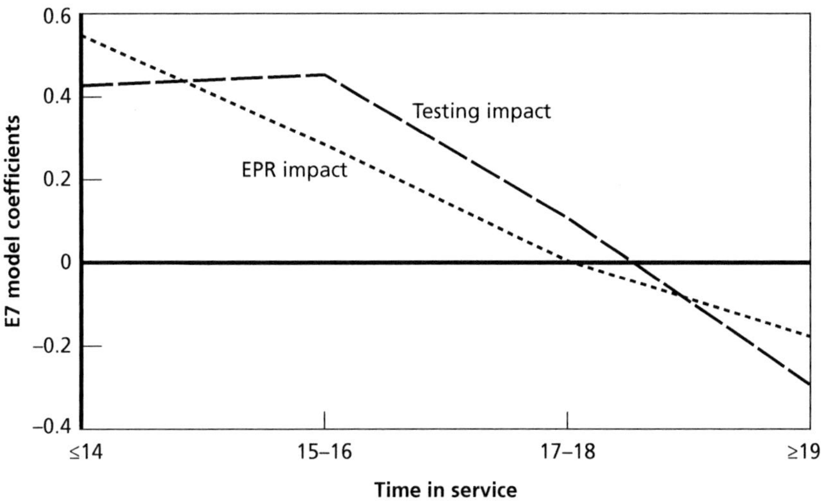

SOURCE: Derived from WAPS history file.
RAND *MG678-H.2*

APPENDIX I

ACT, SAT, and ASVAB Approaches to Standardization

How Are Large-Scale, Commonly Used Tests Standardized?

A number of methods can be used to standardize test scores. These methods, usually called equating methods, are based on statistical concepts. We discuss the methods used on the ACT, SAT I Reasoning Test, and the ASVAB. The developers of each of these tests have successfully used standardization methods to ensure comparability of test scores among the various versions and administrations of its test.

Standardization of ACT Scores

The ACT employs design constraints as well as equating to ensure that ACT test scores are comparable across its different forms and administrations.

ACT employs strict development specifications regarding the types of questions that can be asked. Test design specifications are also used to control the difficulty and content of the questions on each test. Candidate questions are pre-tested with actual students before inclusion on an ACT.

When ACT designs a new test, several forms of the test are developed and all are equated to a base form. The base form is a test from a prior year that has not been released to the public. Currently, ACT equates its test forms to a 1989 test.

ACT uses a true score methodology to achieve comparable test scores. This concept is based on the principle that the ACT is given to randomly equivalent groups of students. In randomly equivalent groups, there is a high probability that the distribution of scores for each group will be the same, so ACT relies on this principle. That is, ACT assumes that the ability level of each group should be the same since the groups are, by design, randomly equivalent. The distribution of the scores should be the same for all the groups because of the randomly equivalent property.

To accomplish equating, ACT administers a different form of the test to each group of students. Each form of the test varies slightly. ACT first computes the raw score for each test taker. The raw score is the number of questions answered correctly. ACT then converts the raw score to a scaled score that makes the distribution of the scores the same for all the groups. That is, in the converted scores, the same numbers of people from each group get each scaled score.

ACT's equating method is most clearly demonstrated by an illustration. Supposed ACT administers Form A, Form B, and Form C of a test to three groups of students. ACT's equating method works by selecting a level such as the percentage of people scoring below 20 on Form A. The raw scores from Form A are scaled (multiplied by a constant), so that the proportion of people scoring below 20 is the same as for Form B. ACT repeats this process and multiplies the scores of the Group C test takers by a possibly different constant so that the same percentage of people in Group C score below 20. All the scaled scores are normed to the base reference form, which is currently the 1989 reference test. This method is repeated for each possible score. The result is a look-up table for each test form, in which the raw score can be converted into a scaled score. The tables now link the scores for the three forms of the test, and the scaled scores are directly comparable to each other and to the reference base.

To prevent error drifts over time, ACT performs equating once a year with a test from a previous year that has not been publicly released. Hence, ACT's equating method keeps scores comparable from year to year. Thus, any specific score on the ACT years ago indicates the same level of ability as that same specific score this year and for all the years in between.

Standardization of SAT I Reasoning Test Scores

The SAT I Reasoning Test is given seven times between October and June of each school year. In addition, different forms of the test are commonly given during each administration. The test has two parts, verbal and math. Each part has multiple sections focused on different aspects of verbal and math abilities.[1] The Educational Testing Service (ETS), which develops the test, employs stringent content and statistical specifications to control the content and difficulty of the various forms of each test. In addition, ETS applies a score standardization process to ensure that the scores on SAT I Reasoning Test are comparable across its different forms and administrations.

To apply its score standardization methodology, ETS computes a formula score for the verbal and math parts of the exam. The formula score is computed by allotting one point for each question answered correctly and forming a sum. From that sum, a fraction of a point is subtracted for each question answered incorrectly. The fraction varies for each question according to the number of choices the test taker had to choose from to answer the question. ETS does not add or subtract any amount for those questions the test taker did not answer, and ETS does not subtract any amount for non–multiple choice questions that are answered incorrectly.

ETS uses a data collection method to link a new test form to previous forms that have already been equated and scaled. ETS appends a section, called an "anchor test" to each new form it administers. The anchor test is a section of a test that has been administered in the past. Hence, the anchor test is a section that appears on the new form and on a previous form, and it serves as the link between the new form and previous SAT I reasoning tests. The anchor test is not counted in a test taker's performance on the test and is not identifiable to a test taker. The sole purpose of the anchor test is to allow ETS to equate the new form to previous forms.

---

[1] ETS introduced a new form of the SAT I Reasoning Test in March 2005. The equating procedure described below applies to the pre–March 2005 SAT I Reasoning Tests. Though a modified process may apply to the new SAT I to account for changes in content and format, the ETS approach to score standardization is the same.

ETS begins its equating process by constructing paired samples of test takers. The first sample consists of test takers who took the new form with a particular anchor test. The second sample that is paired with the first sample consists of test takers who took a previous form of the test that included the same anchor test. The samples are chosen to be representative of the SAT I test taking population. ETS then uses the formula scores on the anchor test common to both samples to adjust for any differences in ability levels between the two sample groups of test takers. Next, ETS uses linear models, equipercentile models, and item response theoretical approaches to arrive at equating functions that relate the formula scores on the new form to formula scores on the previous form. This step gives ETS a way to equate the difficulty of the new form to the previous form after adjusting for any differences in ability level of the two sample groups. The exact method applied to each equating effort depends on which is deemed most appropriate for that equating effort.[2] The result of applying the equating functions is an equated formula score for each test taker in the two samples. Finally, ETS converts the equated formula scores to the familiar scaled scores reported to the test takers and test performance users. The conversion is achieved by applying a mathematical formula to the equated formula scores that results in score values between 200 and 800 in 10 point increments. These scaled scores are thus comparable to previous scores.

Standardization of ASVAB Scores

Nine subtests compose the ASVAB. Each subtest measures test-taker aptitude in one of several areas including mathematical, verbal, technical, and spatial dimensions. Each military service uses a different subset of the subtests and can use different performance measures to emphasize aptitudes especially suited to its needs. These performance measures are called *composites*. To ensure that the composites are consistent over time, the Defense Manpower Data Center standardizes the ASVAB scores.

[2] For example, ETS is vigilant about the effects of equating on gender and race.

The reference base test for the current ASVAB is the Profile of American Youth 1997 (PAY97) study. The PAY97 used a sample of approximately 6,000 American young adults, aged 18 to 23 in 1997, with oversamples of black and Hispanic populations. These PAY97 participants took the computerized adaptive version of the ASVAB under standardized conditions in the third and fourth quarters of 1997. This computerized adaptive version of the ASVAB (Form 04D) employs adaptive Bayesian item selection and scoring rules based on item response theory. Although the PAY97 occurred in 1997, DoD did not adopt the new reference base until 2004.

Prior to using the PAY97 reference base, the ASVAB used data from a 1980 study as its reference base. Data for the 1980 reference base were collected via Form 8A. Form 04D of the current ASVAB was linked to the 1980 reference base via equipercentile equating. The equating between these two forms used a sample of more than fifteen thousand test takers. Test forms were randomly assigned to the test takers in the sample, and each Form 04D subtest distribution was equated to the analogous subtest on Form 8A distribution using an equipercentile-based procedure which equated the means and standard deviations of the two distributions. This procedure resulted in a score conversion that placed the Form 04D scores on the Form 8A score scale.[3]

Form 04D is only one form of the current ASVAB. There are many other forms; to complete the standardization of ASVAB scoring, the other forms had to be equated to Form 04D, the computerized adaptive version of the test. In the past, the various forms of the ASVAB have been equated using an equipercentile procedure for each subtest. In this case, equipercentile equating was determined not to be applicable because the Defense Manpower Data Center began using an item response theory (IRT) scoring method for the ASVAB in 2002. The IRT method was adopted for Forms 25A, 25B, 26A, and 26B. In the IRT approach, ability measures are based on how the test taker responds, and the responses are already on a common metric that is based on large samples. This fundamental characteristic of IRT

---

[3] For details on equipercentile equating, see Segall (1997), pp. 181–198.

makes computerized adaptive testing possible and allows IRT scores from different forms to be treated as interchangeable without further equating.

Although, in theory, no further equating was necessary to equate Form 04D to Forms 25A, 25B, 26A, and 26B, the Defense Manpower Data Center concluded that differences in the data collection mechanism, calibration inconsistencies, variances in motivation among test takers, the different test durations of forms, and other factors all pointed to a need to equate the forms. A linear method that matched the first two moments of the distribution of the estimated ability parameter was used to equate the different forms. Post-equating analysis on the composite scores showed that linear equating resulted in similar qualification rates across all forms of the PAY97 reference base test and the Form 8A reference base tests. In addition, post-equating DoD Manpower Data Center studies showed that the composition of qualified applicants did not differ significantly either in the racial makeup dimension or in the gender dimension. Hence, linear equating appears to be compatible with the IRT-based testing approach used in the ASVAB.

Methods Compared

Major testing organizations use various methods to standardize their test scores. The ACT uses true score methodology to equate sample distributions to achieve comparable test scores. The SAT I uses an equipercentile equating approach but evaluates the appropriateness of applying linear and IRT-based equating methods as well. The ASVAB uses a linear equating method. Each method has been extensively evaluated by each testing organization and shown to be an effective score standardization method for its particular test. Each method is applicable to tests with certain characteristics, features, and media. For example, equipercentile equating is more comprehensive than linear equating, but linear equating is especially useful in cases where the distributions have similar shapes—as in the ASVAB case where IRT testing is used.

References

Department of Defense, Department of Defense Directive 1304.20, *Enlisted Personnel Management System,* December 19, 1984. As of September 9, 2007: http://biotech.law.lsu.edu/blaw/dodd/corres/pdf/d130420_121984/d130420p.pdf

———, Department of Defense Directive 1304.20, *Enlisted Personnel Management System,* July 28, 2005. As of September 9, 2007: http://www.dtic.mil/whs/directives/corres/pdf/130420p.pdf

Duncan, Robert E., An Approach for Equalizing Test Scores for SKT-Exempt AFSCs, Brooks AFB, Texas: Armstrong Laboratory, AL/HR-TP-1994-0020, July 1994.

Fuller, Michael, "Wall-to-Wall Look at African-American Promotions," unpublished Air Force Personnel Center briefing, August, 2001.

Galway, Lionel A., Richard J. Buddin, Michael R. Thirtle, Peter S. H. Ellis, and Judith D. Mele, *Understrength Air Force Officer Career Fields: A Force Management Approach,* Santa Monica, Calif.: RAND Corporation, MG-131-AF, 2005. As of September 9, 2007:
http://www.rand.org/pubs/monographs/MG131/

Hall, Francis J., and Clark K. Nelsen, *A Historical Perspective of the United States Air Force Enlisted Personnel Promotion Policy (1947–1980),* Wright-Patterson Air Force Base, Ohio: Air Force Institute of Technology, LSSR-53-80, June 1980.

Moore, Rick, "Impact Of Senior Rater Endorsements On Enlisted Promotions," unpublished Air Force Personnel Center talking paper, November, 1998.

Los Angeles, The Official Web Site of The City of Los Angeles, homepage, January, 2006. As of November 13, 2006:
http://www.cityofla.org

RAW—*See* U.S. Air Force, Air Force Personnel Center Retrieval Application Website.

Schiefer, Michael A., Albert Robbert, Lionel Galway, Richard Stanton, and Christine San, *Air Force Enlisted Force Management: System Interactions and Synchronization Strategies,* Santa Monica, Calif.: RAND Corporation,

MG-540-AF, 2007. As of September 9, 2007:
http://www.rand.org/pubs/monographs/MG540

Segall, Daniel O., "Equating the CAT-ASVAB," in W. A. Sands, B. K. Waters, and J. R. McBride (Eds.), *Computerized Adaptive Testing: From Inquiry to Operation,* Washington, D.C.: American Psychological Association, 1997.

Shore, C. Wayne, and R. Bruce Gould, *Revalidation of WAPS and SNCOPP,* Volume I: *Background, Procedure, and Statistical Results,* San Antonio, Tex.: Operational Technologies Corporation, October 2004.

U.S. Air Force, *USAF Personnel Plan,* Volume III, *Total Objective Plan for Career Airmen Personnel (TOPCAP),* September 12, 1975.

———, Air Force Policy Directive 36-25, *Military Promotion and Demotion,* June 21, 1993.

———, Air Force Instruction 36-2502, *Airman Promotion Program,* August 6, 2002.

———, Air Force Instruction 36-2605, *Air Force Military Personnel Testing System,* November 14, 2003.

———, Air Force Instruction 38-201, *Determining Manpower Requirements,* December 30, 2003.

———, Air Force Instruction 36-2618, *The Enlisted Force Structure,* December 1, 2004a.

———, Air Force Management Agency, *Air Force Enlisted Grades Allocation Process Handbook,* May 27, 2004b.

———, Air Force Manual 36-2108, *Enlisted Classification,* October 31, 2004c.

———, Air Force Pamphlet 36-2241, Volume 1, *Promotion Fitness Exam (PFE) Study Guide,* July 1, 2005a.

———, Air Force Personnel Center Enlisted Promotions. As of October 25, 2007b:
http://ask.afpc.randolph.af.mil/Docs/EProm/prh.xls

———, Air Force Personnel Center Interactive Demographic Analysis System (IDEAS). As of October 25, 2007a:
http://w11.afpc.randolph.af.mil/vbin/broker8.exe?_program=ideas.IDEAS_default.sas&_service=vpool1&_debug=0

———, Air Force Personnel Center Static Reports. As of October 25, 2007b:
http://wwa.afpc.randolph.af.mil/demographics/ReportSearch.asp

———, Air Force Personnel Center Retrieval Application Website. As of October 25, 2007c:
https://wwa.afpc.randolph.af.mil/AFPCSecure/Default.asp

———, Air Force Personnel Center, WAPS history file. Electronic database, not available to the general public.

Voegtle, Trenda, "SMSgt/CMSgt Chronic Critical Shortage Skills (CCS) Program," unpublished HQ USAF/A1PPP talking paper, August 2006.

WAPS history file—*See* U.S. Air Force, Air Force Personnel Center.